Plant Biotechnology: The Genetic Manipulation of Plants

Plant Biotechnology: The Genetic Manipulation of Plants

Ava Gibson

STATES
ACADEMIC PRESS
www.statesacademicpress.com

Plant Biotechnology: The Genetic Manipulation of Plants
Ava Gibson
ISBN: 978-1-63989-419-2 (Hardback)

STATES
ACADEMIC PRESS

Published by States Academic Press,
109 South 5th Street,
Brooklyn, NY 11249, USA

Cataloging-in-Publication Data

Plant biotechnology : the genetic manipulation of plants / Ava Gibson.
p. cm.
Includes bibliographical references and index.
ISBN 978-1-63989-419-2
1. Plant biotechnology. 2. Plant genetic engineering. 3. Agricultural biotechnology.
4. Transgenic plants. I. Gibson, Ava.
SB106.B56 P53 2022
631.523 3--dc23

Table of Contents

Preface

Plant biotechnology refers to the application of various principles and methods of life sciences to create improved varieties of plants. Cross breeding and mutation breeding are the most widely used practices studied under it. Modern plant biotechnology can be classified into two major fields, namely, plant tissue culture and recombinant DNA technology. In plant tissue culture, the excised part of a living plant, also known as explant, is grown in a sterile plant tissue culture medium. The joining together of DNA molecules from two different species takes place within recombinant DNA technology. Some of the other techniques studied under it are micropropagation and genetic engineering. This book presents the complex subject of plant biotechnology in the most comprehensible and easy to understand language. Different approaches, evaluations and methodologies and advanced studies on plant biotechnology have been included herein. This book will serve as a source of knowledge to a wide spectrum of readers.

A detailed account of the significant topics covered in this book is provided below:

Chapter 1- Plant biotechnology can be defined as the use of tissue culture and genetic engineering to produce genetically modified plants for specific needs and opportunities. Agricultural biotechnology and plant genetic engineering fall under its domain. This chapter will introduce the reader to the various aspects of plant biotechnology.

Chapter 2- The culturing of plant seeds, organs, explants, tissues, cells or protoplasts on chemically defined synthetic nutrient media under sterile and controlled conditions is known as plant tissue culture. There are various methods of plant tissue culture such as seed culture, embryo culture, callus culture, organ culture, etc. This chapter discusses plant tissue culture and its different types in detail.

Chapter 3- Plant cell culture is a collection of techniques used to grow and maintain plant cells under sterile conditions on a nutrient medium. Protoplast fusion is a physical phenomenon in which two or more protoplasts of living cells come in contact and fuse together. The diverse aspects of cell culture and protoplast fusion have been thoroughly discussed in this chapter.

Chapter 4- Micropropagation is the artificial process of producing plants vegetatively through tissue or cell culture techniques. The process by which plant embryos form and develop is called plant embryogenesis. The artificial replication of plant embryogenesis is known as somatic embryogenesis. This chapter has been carefully written to provide an easy understanding of micropropagation and plant embryogenesis.

Chapter 5- Genetically modified plants are those plants which have been genetically altered using various methods like protoplast isolation, somatic hybridization, gene gun method, electroporation, etc. for the purpose of introducing favorable traits. This chapter deals with genetically modified plants and the various methods used in a comprehensive manner.

I would like to make a special mention of my publisher who considered me worthy of this opportunity and also supported me throughout the process. I would also like to thank the editing team at the back-end who extended their help whenever required.

Ava Gibson

1

An Introduction to Plant Biotechnology

Plant biotechnology can be defined as the use of tissue culture and genetic engineering to produce genetically modified plants for specific needs and opportunities. Agricultural biotechnology and plant genetic engineering fall under its domain. This chapter will introduce the reader to the various aspects of plant biotechnology.

Plant biotechnology may be defined as the application of knowledge obtained from study of the life sciences to create technological improvements in plant species. By this very broad definition, plant biotechnology has been conducted for more than ten thousand years.

Plant biotechnology can make two direct contributions to the agricultural industry; it can improve processes to produce the existing type of products and it can lead to new types of products. This is true for industries involved in both the input and output side of agriculture.

Serving the input side of agriculture are the industries involved in the production of plant protection products, seeds and fertilisers. In the field of plant protection, biotechnology can give rise to recombinant microorganisms, such as Bacillus thuringiensis, which can act as biological protection agents against pests and diseases. Additionally, biotechnology can contribute to the development of conventional plant protection chemicals by the identification and cloning of genes involved in the production of enzymes associated with actual or potential target sites for chemical action. This in turn can lead to new knowledge and potential new screening mechanisms. Similarly, monoclonal antibodies can be used as diagnostics for the presence of disease and lead to more intelligent and effective spraying regimes. Perhaps the most important carrier of the effects of the new biotechnology to the market place will be the seed industry. This is not surprising since the major part of the 'new biotechnology' is involved in methods of isolating, modifying and transferring genes. The final input is plant nutrition. Some impact has been made in this area through the selection, production and sale of improved strains of rhizobia for inoculating legumes. However, the originally much trumpeted possibility of nitrogen-fixing cereals, despite a recent surge of interest, remains as far off from the market place as it was fifteen years ago.

Plants are the raw materials for many industrial processes. The introduction of novel traits into plants has the potential to improve many of the industrial processes in which they are used. This is true both for the food and feed industries as well as for the industrial uses of plants. So far not many of the potentials have been realised but, nevertheless, some of the first products in this area are moving towards the market.

Plant Biotechnology in the Seed Industry

Plant breeding was profoundly influenced by the findings of Mendel and the development of the science of genetics. Over the past 90 years it has been immensely successful. But, despite this success, there are a number of limitations.

Conventional breeding cannot introduce traits that are not present, or which cannot be generated by mutation and selection, in the species in question or in species with which it can be crossed. The genetic basis of most desired traits is unknown. The genetic map of maize at the beginning of the 1980s was based mainly on mutations and traits of no conceivable commercial importance. In the main, the plant breeder has to infer the genotype from the phenotype; this inference is greatly confounded by the large interaction with the environment. Screening technologies can be expensive and tedious. Thus there is a considerable potential for plant biotechnology to contribute novel and better solutions. Figure represents the type of decision tree that a generalised plant breeder could face in trying to improve his crop for a given trait. At many stages the breeder might be faced with a difficult or impossible problem, in some cases biotechnology offers a potential solution. This is particularly true where the variation required does not exist within the species, or its sexually compatible relatives, but does exist somewhere else in the biological kingdom. For example, the inability of many crops to withstand damage caused by Lepidopteran pests can be combatted by transferring the gene(s) for delta-endotoxin synthesis present in the bacterium Bacillus thuringiensis. Demonstrated field efficiency of this transfer has been achieved in tobacco, tomato and cotton.

Many of the problems of understanding the genetics of important traits is and will continue to be improved by using RFLP technology. This technology will aid tremendously in incorporating traits that are controlled by several genes into elite material. RFLP technology also can and will enable breeders to backcross simply inherited traits from exotic material much more effectively. The problem in the past has not been to incorporate the traits but rather to eliminate the phenomenon of'linkage drag'. This brings along with the favourable trait large proportions of genetic material which are linked to that trait. If this linked genetic material does not have any readily identifiable phenotypic markers it is difficult to select against. However, leaving it in the elite line may result in lowered performance of the elite line. Retrospective studies by Young & Tanksley have shown that large amounts of linked genetic material from the donor line have not been eliminated in many backcrossing programmes.

It is one thing to identify where plant biotechnology may be useful in a breeding programme, it is something entirely different to decide if using the technology will pay off commercially. In the long term, industry will not be interested in investing in plant breeding and biotechnology unless it allows industry to give the people who invest in it a return on that investment. To date there has been virtually no return on investment in plant biotechnology despite the vast amount of funds invested in it. One consulting company has calculated that about US$2.5 billion have so far been invested by multinationals and small start-up companies. In many cases small firms have been unable to sustain the endeavour and have had repeatedly to search for new funds. These have not always been forthcoming and the companies have either folded or merged with others. Likewise, many major company investors have left the field. In particular, most of the major oil companies have exited. Others, particularly agriculturally based companies, appear to have a longer-term commitment. However, it is unreasonable to expect that this will continue indefinitely without adequate returns being generated.

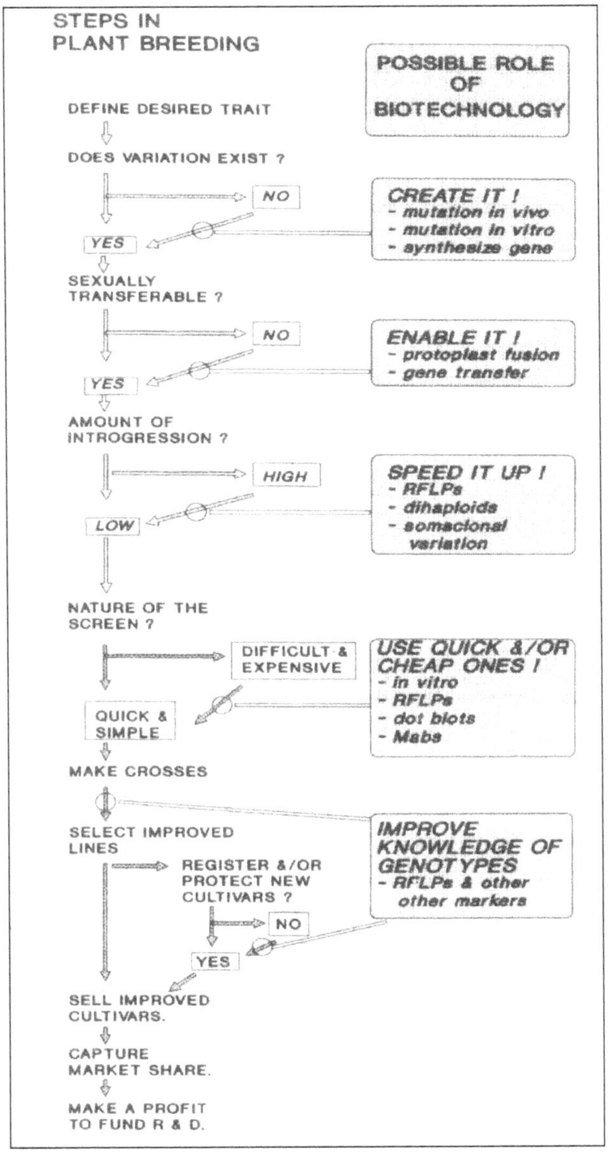

Figure: Decision tree for a plant breeder trying to improve a crop for a given trait and points at which biotechnology may make an impact.

A measure of disillusion with plant biotechnology and plant science in general has also been seen in the public sector. This has probably been most dramatic in England where the AFRC has gone through a large cut in resources. The Plant Breeding Institute at Cambridge, once the centre of the AFRC's effort and the premier institution combining breeding and plant biotechnology in the world (either public or private), has been divided up and the breeding part sold. In the U.S.A., critics who published the Biotechnology's Bitter Harvest petition politicians to reduce public research spending on certain areas of plant biotechnology. Individual plant scientists have also written public statements criticising the spending of public money on this field and also criticised the role of scientists in facilitating the development and application of the technology to agriculture.

This critical social and financial environment threatens investment in both public and private plant science and technology. Furthermore, if this technology brings the benefits expected, limiting investment could severely limit society's efforts to feed an ever-expanding population.

Science into Practice

Figure summarises the movement of science into practice in society. Science per se is concerned with discovery but with an implied belief that knowledge brings potential benefits. Many commentators on science, for example Medewar (1984), see science as being carried out by ordinary people motivated by curiosity and a desire to make the world a better place to live in. Science can in general only be turned into benefits through the development of technology (i.e. the development of new tools and products). Once developed, technological innovations have to be implemented and accepted by society. To be accepted, they need to bring perceived benefits to society and, if society wishes to have a flow of new products from technology, a way has to be achieved to give reward to those that carry out the research and development. The mechanisms by which this may be achieved may differ between the private and public sector but the underlying principle is the same. Besides financial returns, there must be political support for the technology, because without this it is difficult for the champions of the technology to get new projects off the ground, either in industry or in the public sector. The whole chain of events also cannot occur unless there is a cadre of dedicated, trained and able people to carry it out. The motivation of young people to enter the field is usually vitally affected by the esteem in which it is held by the public. Thus the path of development from a scientific discovery to the arrival of new useful products in the market place is subject to a range of feedbacks.

The way in which these feedbacks occur is also shown in Figure. The roles of the public and private sectors differ in their responsibilities for the different parts of the chain. Basic science is seen as the province of the public sector, although industry does make a contribution. The development of technology and the making of inventions occur in both areas, although, despite the increasing number of patents being applied for by public sector institutions, the majority of technology inventions come from industry. Where industry plays the largest part is in the implementation of inventions. Without implementation there is no innovation and inventions stay on the drawing board or in the patent literature. It is this area where the Japanese have been particularly strong and have brought to the market many inventions made elsewhere (e.g. the transistor). In the seed industry, the public sector in Europe and the U.S.A. has largely withdrawn from producing finished varieties - the essential final step for implementing most plant biotechnology. Only in Japan, where the seed industry is largely government controlled, and in the institutes of the CGIAR are there large-scale publically funded efforts to implement the inventions of plant breeding and biotechnology. The conclusion is that in Europe and North America at least, the products of the genetic revolution in plants will chiefly reach the market via private industry.

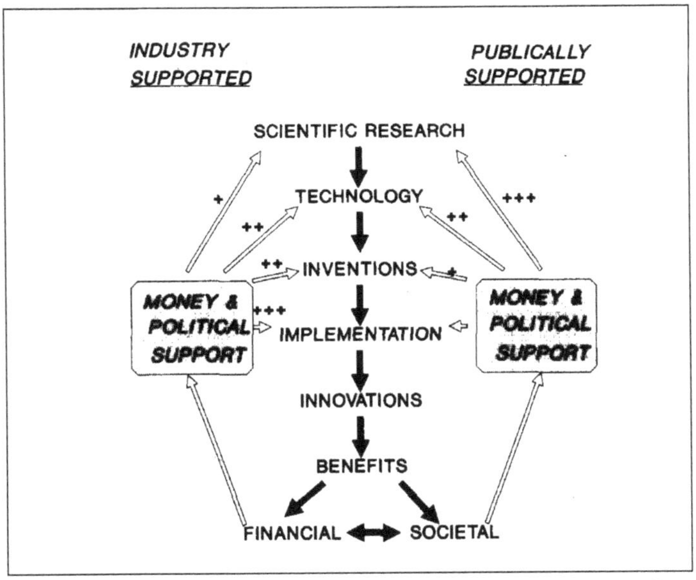

Figure: The path of discoveries from scientific research to the marketplace and feedback controls operating upon it.

The Blocks on Implementation

There are three major areas which have to be satisfied in ensuring the implementation of biotechnology by industry; financial benefits, a rational regulatory environment and public acceptance. The latter two areas also apply to introduction of products from the public sector.

Financial Benefits

These are affected by the structure of the industry and the way in which inventions are protected (so-called intellectual property protection). For example, in the European seed business the return to the breeder of inbred varieties in Europe is through a royalty system. The overall royalty income available to breeders in the U.K. has been calculated as being about half the amount of money currently invested in the commercial breeding programmes in the U.K. Obviously, this situation cannot continue indefinitely. The farming industry and the society depending upon it has a choice between working out a way to ensure that the breeders receive a fair return on their efforts or eventually losing the value of the breeding effort that is currently improving cereal varieties. For example, Pioneer recently chose to close down a large part of their breeding programme for wheat in the U.S.A. because of insufficient return. Additionally, if these crops are to benefit from the efforts of biotechnology then clearly a means to give even more return to the inventor will have to be devised - the alternative is that companies will cease to invest in technology improvement for inbred cultivars.

The ownership of varieties is determined through the various national systems on plant variety rights. In essence, these grant a breeder an intellectual property right for his new variety; other breeders may not obtain rights for the same or nearly identical varieties. This ensures that the royalties go to the breeder of a registered variety and that others cannot multiply that variety and claim it as their own. Patents, another intellectual property system, also work to give the inventor a fair return and to stimulate new inventions. In both cases the interests of society are supported by a system that stimulates investments to provide improved solutions to agricultural problems. Despite many criticisms, the patent system has worked well to ensure the development of new products for society.

Regulations

For the benefits of rDNA technology to come to the market place, the products must be tested and developed under current regulations and new regulations governing commercial release must be developed. Field testing of rDNA plants has occurred widely in Europe and the U.S.A. (with some 300 field tests to date) although there are some European countries where regulations are not yet in place or in which few tests have been held. So far these regulations seem to work well and do not provide any serious block to the development of rDNA plant products.

More importantly, if rDNA plants are to come to the market place, there is the need to develop regulations that govern the commercial release of products. It is essential that any regulations produced are rationally based and make use of our existing body of knowledge. For example, recent proposed regulations governing the labelling of food would require products of inter-species gene transfer to be labelled as the product of genetic engineering. Many foods are already the result of inter-species gene transfer or of inter-species hybridisation. Take the case of wheat flour. Wheat is a hexaploid derived from a chance natural association of three genomes from different species. Furthermore, around half of the wheat grown today in Britain is derived from lines in which the short arm of chromosome 1 of the wheat B genome has been replaced by the short arm of chromosome 1 of rye (Secale cereale, a different genus). The advantage of this translocation is that it introduces genes for disease resistance and higher yield. It also introduces genes for 'sticky dough' which limits the use of these varieties for bread making. Neither the flour from this wheat, nor any products made from it have to be specially labelled under current or proposed legislation. However, if the recently published draft proposed guidelines for the U.K. are followed, then cultivars derived from the transfer of the defined favourable genes from rye through the isolation of those genes and their subsequent transformation into wheat will have to be labelled. This discrimination in favour of the crude transfer against the precise transfer does not seem rational. Much more important is that people should know what is in a product and if it differs in any way from what is traditional. However, the attention should focus on the product of the gene (i.e. protein or enzyme reaction) and not on the gene or on how it was introduced.

Public Perception

If the products of plant molecular biology are to make a contribution to the market place, then they have to be accepted by the public. This means that their benefits have to outweigh any real or perceived risk. This acceptance is affected by the public's ability to judge and on the activities of critics, proponents and the media.

1. Benefits: Breeding has for the last ninety years been a story of the increasing precision and control over recombining genes in our crop plants to give increased performance. The result of this effort in our major crops has been a steady increase in yield potential. However, despite this success, many problems remain as indicated above particularly in handling the large number of genes in crop plants and in having sufficient variation available within the species. This difficulty can be considered in relation to having to protect crops against attacks from pests and diseases. Certain crop species are partic-ularly susceptible to given pests and diseases. For example, cotton is heavily attacked by insects and requires large amounts of insecticides in order that it can be grown profitably in the U.S.A.; in contrast, sorghum and soybeans, even when grown in the same environment, require relatively little. If sources of pest tolerance exist in sexually crossable species these can then be bred into the crop but in many cases such sources of resistance are not present in the available gene pool. In this instance, the breeders' art can contribute little to the solution of the problem.

rDNA technology offers a contribution by its ability to isolate and transfer genes by means not limited by sexual compatibility. This means that a whole new range of variation is available to plant breeders. Thus the problem in cotton can at least be partly overcome by introducing genes producing proteins toxic to the insect. Such modified cotton plants are effective in combatting insect attack. Adding such remedies to the plant has several advantages in ease of use for the farmer and in ensuring the remedy is constrained, so that it is only exposed to the target pest and not to other insects in the environment that do not feed off the crop plant.

Overall, the remedies provided via plant breeding, aided by molecular biology, increase the possibilities available for solving the problems surrounding agriculture and the environment, and have the potential to enable man to continue to improve his crop plants and provide more food. In situations, as in Europe, where the amount of total food production is sufficient or in surplus, crop improvement through these genetic technologies can contribute to better and more consistent yields and more effective use of resources (land, fertiliser, energy etc.). The increase in solutions provides greater choice for the farmer, the regulator and society in general. If the choice is exercised wisely, then the extra options provided by genetic technologies should allow agriculture to become more environmentally compatible and sustainable.

2. Risks: At the inception of rDNA technology, it was realised that it was a powerful technology that should be developed carefully. Consequently, following the Asilomar conference in 1973, a series of guidelines were set up in the U.S.A. and other coun-tries to regulate these activities. This was prudent and such prudence is still required in certain areas of the technology. However, in the vast majority of cases additional risk can neither be perceived nor measured. For plants, the US National Academy of Sciences have concluded that "the risks associated with the introduction of rDNA-en-gineered organisms are the same in kind as those associated with the introduction of unmodified organisms and organisms modified by other methods." Nevertheless, giv-en the current level of public anxiety, it seems prudent to continue with a step-by-step, case-by-case approach to the release of transformed plants into the environment. In addition, data is being gathered in a number of experiments to evaluate the potential spread of genes out of crop plants into plants growing in the natural environment. This will enable one of the major concerns - that of the escape of introduced genes into wild species - to be addressed. However, there is little or no evidence that any major genetic trait introduced into crop plants by conventional methods (including natural-ly occurring, herbicide tolerance traits) have either "escaped into weeds" or caused a problem by doing so.

Another concern, that is often raised, is that the introduced gene will cause unforseen and dramatic changes in the properties of the crop. This potential risk is not restricted to rDNA derived plants. Every new genotype created by the plant breeder brings about a novel assortment of genes which may have very profound effects. To ensure that these effects are known, all new genotypes undergo a rigorous programme of testing and selection ending, at least in Europe with a minimum of two years of official, independently-regulated trials. In the same way, rDNA plants will also undergo considerable testing before commercial release to ensure their fitness for use. While it is impossible to say that no unforseen effects will occur, this does not alter anything from the existing situation and potential risks will be dealt with in the same way as in existing plant breeding practice.

3. Public acceptance: A number of factors affect the perception of the public of

the acceptability of a new technology. In general these have led to a low percentage of support for biotechnology in surveys of opinion. For example, surveys in Denmark suggest that only just over 20% of those asked support gene technology with about 65% being neutral. Some of the factors involve include:

(a) anxiety and fear of science and technology; (b) the activities of critics with many different motives; and (c) the apathy or ineptitude of practitioners of molecular biology in stating their case to the public.

Ignorance of science is an integral part of modern society. Assessments of scientific literacy in the population of Western Nations indicate that it is very low and not changing with time. Ignorance breeds fear. Technology also means change, another uncomfortable process to the majority of people. History shows that new ideas, new technologies and change in general take a considerable time to become accepted. It is pointless to criticise or rail against this attitude. Furthermore, the public's natural reticence has been reinforced in recent years by disasters arising from new technologies - particularly nuclear energy. Scientists and technologists no longer have the trust and reverence they once had. My conclusion is that those who believe in the beneficial balance of plant molecular biology and biotechnology for society at large must make their case to that society. In doing so they must address the genuine fears that society holds for new technologies.

The general public is also being addressed by a range of critics of biotechnology. To win the battle with these critics over the benefits and risks of technology, it is important to understand their motivation. Rarely is this based on strict evaluation of the facts. Firstly, criticism is in itself a business. Large organisations have built up to combat industry and governments. Irrespective of one's opinion of the rights and wrongs of a particular issue, these organisations have the need to perpetuate themselves, to raise funds and to be financially viable. To do this they may target areas that are tangential to their real objectives but which serve to grab the attention and support of the public. Plant biotechnology is a highly visible area of agriculture - it therefore provides a 'hook' which can be used to attract attention to other issues. At least some of the critics appear to have as their 'real' target the structure of agriculture in both developed and developing countries and the role of multinational organisations. Although biotechnology probably has little impact on this structure it does provide a platform from which to raise the issue.

Critics are also motivated by political power, particularly the desire to be elected or re-elected. This can be seen in the rise of the Green Party in Germany or in the 'conversion' of the Conservative Government in the U.K., to environmental issues in the late 1980s. In many cases the critical attitude has not arisen by conviction or rational conversion but by the appreciation that such a stance might give an electoral advantage.

The attention of the public is a powerful ego-enhancing drug. Being a critic and portrayer of doom and uncertainty is much more likely to attract attention than a rational presentation of benefits and risks. Such critics are not to be converted by rational arguments - they trade on the emotions of the public and their ego is enhanced by the attention of the media. Bad news and anxiety makes good headlines and sells newspapers - careful explanations of technology and a balanced approach to problems only has limited media value. It is probably impossible and irrelevant to involve oneself in a rational debate of the issues directly with critics motivated in this way.

Many problems such as the structure of agriculture, the economics of the developing world, the growth of the world's population, are so difficult as almost to defy analysis and solution. Nevertheless, many people are desperate to propose a solution and it becomes attractive to grab at easy answers without full consideration as to whether they would help, hinder or be irrelevant. In such circumstances, it is also more comfortable to place the blame elsewhere. In the western world, industry is a convenient 'fall guy'. Under such circumstances plant biotechnology and the industrial firms that practice it are easy targets. However, if biotechnology was de-invented or stopped in its tracks it would not change the problems - they would still exist. It is also difficult to come up with a convincing alternative to the current mixed economic system of the western world. Industry in the western world is in general owned by and responsive to a large number of people and develops in relation to social pressures and the demands of the market. This market is made up of individuals, critics and proponents included, and whether we individually approve of a technology or development or not, it will only proceed if it receives sufficient support of the public. This suggests that the arguments of such critics can be met by encouraging them to face the real and difficult issues underlying some of the above problems.

Because of the issues discussed above, the lack of positive support for biotechnology that has been found in various surveys is not surprising. The position is not improved when scientists make no attempt to explain the technology and its advantages to the public. Many industries have adopted a low profile hoping that the problem will go away. Insufficient knowledge of the science and technology has reached into schools. Where attempts have been made to bring the technology and the benefits that it might bring to the notice of the general public the results have been encouraging. This effort needs to be intensified by both the public and private sector. In other words it is up to all people knowledgeable in the technology, and who believe it will benefit society, to put this view over to the public. Probably the best place to start is in the education system with teachers and pupils. Ingram (in this volume) has made some suggestions how this may be done. Other models exist, but liaison between local education authorities and schools and scientists willing to help explain the technology is an avenue open to most. Unless scientists take the responsibility to inform the public on the nature of molecular biology and how the advances might be used, support for both the science and products of molecular biology will not be forthcoming.

Plant Products of Biotechnology

Plant products of biotechnology have been available in the market. These modified crops look like their traditional counterparts, but they possess special characteristics that make them better. These crops offer several benefits both farmers and consumers. Farmers gain higher crop yields and have increased flexibility in management practices while consumers have "healthier crops" (i.e., crops grown with fewer pesticides and/or with healthier nutritional characteristics). Plant products of biotechnology approved for food use have been modified to contain traits such as:

- Insect resistance.
- Disease resistance.
- Herbicide tolerance.
- Altered nutritional profile.
- Enhanced storage life.

Examples of Plant Products of Biotechnology

Product	Trait
Alfalfa	Herbicide tolerance, altered lignin production.
Apple	Non-browning
Bean	Virus disease resistance.
Canola	Herbicide tolerance, modified oil/fatty acid, pollination control system, phytase production.
Cotton	Herbicide tolerance, insect resistance, low gossypol.
Eggplant	Insect resistance
Flax, Linseed	Herbicide tolerance
Maize	Abiotic stress tolerance, altered growth/yield, herbicide tolerance, insect resistance, modified product quality (modified alpha amylase, lysine boost, phytase production), pollination control system.
Melon	Delayed ripening
Papaya	Disease resistance
Pineapple	Delayed ripening, modified fruit color
Plum	Disease resistance
Potato	Disease resistance, herbicide tolerance, insect resistance, modified product quality (modified starch, reduced acrylamide potential, non-bruising), fungal disease resistance.
Rice	Herbicide tolerance, insect resistance, anti-allergy, modified product quality (biofortified with Provitamin A, anti-allergy).

Rose	Modified flower color
Safflower	Modified product quality
Soybean	Herbicide tolerance, insect resistance, modified product quality, altered growth/yield, abiotic stress tolerance, modified oil/fatty acid.
Squash	Disease resistance
Sugar beet	Herbicide tolerance
Sugar cane	Insect resistance, drought tolerance
Sweet pepper	Disease resistance
Tomato	Disease resistance, insect resistance, delayed ripening, delayed fruit softening.
Wheat	Herbicide tolerance

Biotech Soybean

Soybean is the oil crop of greatest economic relevance in the world. Its beans contain proportionally more essential amino acids than meat, thus making it one of the most important food crops today. Processed soybeans are important ingredients in many food products.

Herbicide Tolerant Soybean

Herbicide tolerant soybean varieties contain a gene that provides resistance to one of two broad spectrum herbicides. This modified soybean provides better weed control and reduces crop injury. It also improves farm efficiency by optimizing yield, using arable land more efficiently, saving time for the farmer, and increasing the flexibility of crop rotation. It also encourages adoption of no-till farming - an important part of soil conservation practice. These varieties are the same as other soybeans in nutrition, composition, and in the way they are processed into food and feed.

Insect Resistant Soybean

This biotech soybean exhibits resistance to lepidopteran pests through the production of Cry1Ac protein. Insect resistant soybean was developed to reduce or replace high insecticide applications and at the same time maintain soybean yield potential.

Oleic Acid Soybean

This modified soybean contains high levels of oleic acid, a monounsaturated fat. According to health nutritionists, monounsaturated fats are considered "good" fats compared with saturated fats found in beef, pork, cheese, and other dairy products.

Oil processed from these varieties is similar to that of peanut and olive oils. Conventional soybeans have an oleic acid content of 24%. These new varieties have an oleic acid content that exceeds 80%.

Biotech Maize

Maize is one of the three most important grains of the world. It is used as livestock feeds, processed as cooking oil and food additives, and currently as feedstocks for biofuels.

Herbicide-tolerant Maize

These maize varieties work in a similar manner as herbicide-tolerant soybean. They allow growers better flexibility in using certain herbicides to control weeds that can damage crops.

Insect Resistant Maize

This modified maize contains a built-in insecticidal protein from a naturally occurring soil microorganism (Bacillus thuringiensis) that gives maize plants season-long protection from corn borers. This means most farmers do not have to spray insecticide to protect maize from harmful pests, which can cause significant damage and yield loss in many areas. Bt maize also reduces toxin contamination arising from fungal attack on the damaged grain. The Bt protein has been used safely as an organic insect control agent for over 50 years.

Biotech Rice

Rice is life for more than half of humanity. It is the staple food for over 3 billion people, more than 90% of whom are Asians.

Herbicide-tolerant Rice

These rice varieties work in a similar manner to herbicide-tolerant soybean. They contain a gene that provides resistance to one of two broad spectrums, and environmentally benign herbicides.

Biotech Tomato

Delayed-ripening Tomato

The delayed-ripening tomato became the first genetically modified food crop to be produced in a developed country. These tomatoes spend more days on the vine than other tomatoes, thus resulting in better flavor. Furthermore, the longer shelf life has commercial advantages in harvesting and shipping that can reduce the costs of production.

Biotech Cotton

Herbicide-tolerant Cotton

This cotton works in a manner similar to other biotech tolerant crops. For benefits, see herbicide-tolerant soybean.

Insect Resistant Cotton

This modified cotton works in a manner similar to insect resistant corn. It contains a protein that provides the plant with season-long protection from budworms and bollworms. The need for additional insecticide applications for these pests is reduced or eliminated.

Biotech Potato

Insect Resistant Potato

This biotech potato works like insect resistant corn. It contains a protein that provides the plant with built-in protection from the Colorado potato beetle. Thus, this potato needs no additional protection for this pest, benefiting farmers, consumers, and the environment.

Virus Resistant Potato

Several potato varieties have been modified to resist potato leafroll virus (PLRV) and potato virus Y (PVY). In the same way that people get inoculations to prevent disease, these potato varieties are protected through biotechnology from certain viruses. Furthermore, virus resistance often results in reduced insecticide use, which is needed to control insect vectors that transmit viruses.

Low Acrylamide Potato

This biotech potato has 50-75% lower levels of acrylamide, a potential carcinogen in humans, produced when potatoes are exposed to high temperatures. It is also less susceptible to bruising.

Biotech Canola

Canola is a genetic variation of rapeseed and was developed by Canadian plant breeders specifically for its nutritional qualities, particularly its low level of saturated fat.

Herbicide Tolerant Canola

Herbicide tolerant canola contains transgenes conferring tolerance to herbicides. This is similar to the trait exhibited by herbicide tolerant soybean.

High Laureate Canola

These canola varieties contain high levels of laurate. Oil processed from these novel varieties is similar to coconut and palm oils. This new canola oil is being sold to the food industry for use in chocolate candy coatings, coffee whiteners, icings, frostings, and whipped toppings. Benefits extend even to the cosmetics industry.

Biotech Alfalfa

Alfalfa is one of the most important legumes used in agriculture.

Herbicide-tolerant Alfalfa

This alfalfa works in a manner similar to other HT crops.

Biotech Papaya

Virus Resistant Papaya

This Hawaiian-developed papaya contains a viral gene that encodes for the coat protein of papaya ringspot virus (PRSV). This protein provides the papaya plant with built-in protection against PRSV. This biotech papaya works in a manner similar to virus resistant potato.

Biotech Squash

Virus Resistant Squash

A biotech yellow crookneck squash is now able to resist watermelon mosaic virus (WMV) and zucchini yellow mosaic virus (ZYMV). These new varieties contain the coat protein genes of both viruses. This biotech approach bypasses aphid control, which may reduce or eliminate the use of insecticides.

Applications of Plant Biotechnology

Genetic engineering of plants provides an opportunity to alter their properties or performance in order to improve upon their utility. Such technology may be used to modify the expression of genes already present in the plants, or to introduce new genes of other species with which the plant cannot be bred conventionally. Thus, it gives greater efficiency to the fulfilment of conventional breeding purposes.

One of the significant applications of such techniques lies in adding single genes to desirable plant types. Plant transformation can be used to introduce new or novel characteristics that create a new market or displace conventional products. The improvement may relate to the nutritional value of the plant or the functional properties in processing or even consumption per se. Above all, this technology broadens the possibilities of transferring genes between unrelated organisms, and thus creates novel genetic information by specific alteration of cloned genes.

Food Quality

Nutritional Quality

Seed crops play an important role in human and animal nutrition. Just a few cereals contribute to nearly fifty per cent of the total food calories. Similarly, seven species of grain legume account for a large portion of our calorie intake.

However, cereals and legumes contain certain proteins that are deficient in amino acids like lysine and threonine. Legumes are also deficient in sulphur amino acids. Some other seed crops like rice offer a better balance of amino acids, but fall-out on their overall protein levels.

Common logic follows that each of these foods could be catapulted to perfection if their deficiencies could be overcome by borrowing those missing traits from other crops. That is exactly what plant biotechnology does – the transfer of single or multiple genes to plants lacking important components.

Genetic tools can be used to alter the carbohydrate, fat, fibre and vitamin content of food. Another useful application is that of picking up genes from protein-rich cereals and transferring them to low-protein food. Transgenic tools are also being used to improve the nutritional value of crops by reducing their anti-nutritional factors (like protease inhibitors and haemaglutinins in legumes). Problems associated with the flatulence in certain foods can also be addressed by manipulating the dietary fibre and oligosaccharide content.

Biotechnical applications are extremely useful in case of wheat as well. The quality of wheat is determined by the presence of seed-storage proteins of the grain. Thus, its quality may be improved by manipulating the presence of these proteins. More gluten proteins can also be added to give enhanced elasticity to dough. Further, the starch content of wheat can be altered to suit the properties of products like noodles.

Functional Quality

Transformation can be applied to fruits and vegetables to improve their flavour and texture by manipulating their maturing process. The performance of plant products during their processing can also be improved by genetic engineering. For instance, the first genetically engineered food, the Flavr-Savr tomato was genetically manipulated to slow down its ripening, and has a longer shelf-life.

Another common strategy to control ripening is to curb the production of the ripening hormone ethylene. Ethylene is produced from S-adenosylmethionine by conversion to 1-amino-cyclopropane- 1-Carboxylic acid (ACC) in the presence of ACC Synthase, followed by the generation of ethylene by an ACC oxidase or ethylene-forming enzyme.

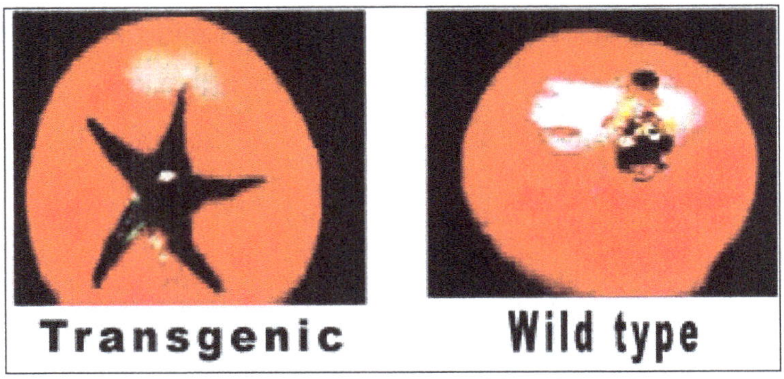

Figure: Transgenic tomato with delayed ripening.

Ripening can be delayed by directing antisense constructs against either of these enzymes, or by removing ACC with an ACC deaminase. Fruits may then be ripened as required by exposure to an artificial source of ethylene.

Malting and Brewing

The production of beer involves the germination of barley under controlled conditions. The quality of beer thus depends largely upon the composition of the barley grain. Many qualities of these grains can be substantially improved through genetic engineering. For instance, improving the stability of the barley enzymes (especially at high temperatures) can enhance its effectiveness at the temperature used during mashing. The flavour of the beer can also be manipulated by genetically treating the barley. One such technique is that of reducing the levels of lipo-oxygenase.

Storage Carbohydrates

Increasing the levels of certain enzymes like ADP pyrophosphorylase can enhance the starch synthesis of food products. This can improve the yields of starchy foods. Transformation can also alter the properties of plant starches. The proportion of amylase and amylopectin in the starch and the quality can also be regulated. This would allow the tailoring of starch to meet requirements for specific foods or industrial products.

Transgenic plants with increased levels of fructans (a form of glucose) are already being produced using a levansucrase from bacteria. The sucrose content of plants can also be manipulated to enhance the quality of sugar crops like sugar cane and sugar beet.

Disease Resistance

Insect Resistance

Genetic engineering has proved to be a boon for producing pest-resistant plants. This technology has overcome the shortcomings of using chemical pesticides. Of late, the technique of introducing disease- resistant genes into plant species has also gained tremendous popularity.

For instance, protease inhibitors can prevent the digestion of proteins by insects, and hence slow down their growth rate. The transfer of such proteins to the plants acts as a natural protection mechanism against insect attack.

Certain bacterial genes have also proved to be quite effective in preventing pest damage. Bacillus thuringiensis (Bt) produces Bt Toxin, which is effective against insect larvae. Transgenic plants harbouring Bt genes have been produced in crops like soya bean, maize and cotton, and have proved to be resistant to pest attacks.

Many other serochemicals (chemicals which alter the insect behaviour) are produced by certain insect and plant species. Transferring these to other plants can be very effective in checking disease incidence. To take another example, the susceptible potato crop does not contain anti-feedant chemicals like farnase, a terpenoid and other related compounds.

These are produced by aphid resistant plant species such as Solanum berthaultii (in leaf hairs). These compounds act by eliciting an attack response in aphids, so that they are unable to establish themselves on the crop. Transferring these genes to the potato crop can protect it from aphid menace.

Virus Resistance

Production of transgenic plants with resistance to viruses is one of the most successful applications of plant transformation. Several strategies involving the expression of the viral genome in the plant have proven effective. For example, the expression of coat protein gene from virus has been widely successful. Both sense and antisense expression of parts of the viral genome can be protective against viral infection.

Nematode Resistance

Novel genes for nematode resistance offer an alternative approach to the production of nematode-resistant plants. Genetic engineering provides an opportunity to develop transgenic plants with genetic resistance to these long-term plant pests, and thus reduce the reliance on chemical nematicides in agriculture.

Herbicide Resistance

The choice of a herbicide is very critical as it carries a high risk of inducing resistance. Weeds may rapidly develop multiple herbicide resistance in some systems when several classes of herbicide act on the same molecular target. Here again, herbicide resistance genes offer protection by detoxifying the herbicide (converting it to an inactive form).

Enhancing Photosynthetic Efficiency

The process of photosynthesis is the most significant mechanism for adding energy to the plants. However, even the most efficient plants can utilise only about three to four per cent of the full sunlight. Biotechnology is now being used to improve the level of photosynthetic efficiency of RuBPCase (Ribulose bis phosphate carboxylase, involved in carbondioxide fixation).

This enhances the efficiency of catalysis and reduces the competitive oxygenase function. Useful variants can also be produced by combining the genes coding for large and small sub-units of the enzymes from different species.

Abiotic Stress Tolerance

Plant productivity suffers major losses due to various forms of stress during the course of their development. These stress factors include temperature, salinity, drought, flooding, UV light, and various infections. While the molecular basis of such responses are not yet clear, we know that they include de novo synthesis of specific proteins (under temperature shock) and enzymes (alcohol dehydrogenase under anaerobiosis and phenyl alanine amino lyase under UV irradiation).

The genes responding to abiotic stress have been cloned and sequenced in many laboratories. The regulatory sequences of some of the genes have also been identified. For instance, the 5′ promoter sequence of alcohol dehyhrogenase has been linked to the CAT reporter (Chloremphenicol Acetyl Transferase) gene and transferred to tobacco protoplasts where O_2-sensitive expression has been demonstrated.

Such environmentally inducible promoters will certainly become useful tools to study gene expression, and this work will lay the foundation for the transfer of stress-responsive genes under regulated promoters to susceptible species. Recently tomato plants that are resistant to salinity have been developed.

Genes from varied organisms like marine resources can also be used to improve upon plants in various ways. This is an innovative step towards developing salt tolerant species, by transferring the genes from marine plants (halophytes) to grain and vegetable crops.

Similarly a gene, which encodes a protein from a flounder fish, has been transformed to plants to protect them against freezing damage. This protein could be useful in preventing frost damage in post-harvest storage. Thus, freezing could be used to preserve the texture and flavour of some fruits and vegetables, which are currently not suitable for freezing.

Development of Nitrogen Fixing Capacity in Non-Leguminous Crops

While the application of nitrogenous fertilisers has proved to be an efficient route to improving crop yields, it continues to be an expensive proposition. The alternative is to provide a natural source of nitrogen within the plant. Introducing nitrogen-fixing microorganisms can do this.

Such micro-organisms are capable of fixing atmospheric nitrogen in the presence of nitrogen fixing bacteria Rhizobium. Transforming the nitrogen fixing genes (nif genes) from leguminous to non-leguminous crops can offer a cost-effective alternative to the expensive fertilisers.

However, other ways of improving the nitrogen yield in plants can be achieved by increasing the efficiency of fixation process in symbiotic bacteria, increasing the efficiency of fixation process in the synthetic bacteria, modifying the nitrogen-fixing bacteria to maintain nitrogen fixation in presence of exogenous nitrogen.

Cytoplasmic Male Sterility

A lot of research has gone into explaining the mechanism of Cytoplasmic Male Sterility (CMS). This trait results in the production of non-functional pollen in mature plant species like sorghum, maize and sugar beet, and hence facilitates the generation of valuable high yielding hybrid seed.

Cytoplasmic male sterility in these plant species is basically associated with the reorganisation of mitochondrial DNA and the synthesis of novel polypeptides. The rapidly developing biotechnological tools may eventually enable the transfer of the CMS trait to male fertile lines. Genetically engineered male sterility also holds a great potential for generation of hybrids in agriculture.

Plant Development

Development of a plant is a complex process, which involves the role of light receptors like phytochrome, chloroplast gene expression, mitochondrial gene expression in relation to male sterility, storage product accumulation, and storage organ (fruits) development.

It is now possible to clone and sequence various genes responsible for plant development. This has increased the possibility of manipulating the expression of these genes, and subsequently the process in which they are involved. For instance, early flowering genes have been reported to alter the properties of the late maturing varieties.

The isolation of specific promoter elements has also helped design crops that express proteins in specific tissues. Genes responsible for colour formation can be transferred to plants bearing colourless flowers. What's more, manipulation of genes that control flowering and pollen formation can generate transgenic plants with altered fertility. Expression of leafy and APETALAI gene in Arabidopsis has resulted in precocious flowering.

Similarly, the putative hormone receptors in plants influence the sensitivity of different tissues to growth regulators, and their subsequent differentiation and development. The introduction of wild type or modified genes for specific growth regulators has proved effective in manipulating plant development (like changing the maturity time or the number and size of potato tubers). This approach can be applied for modifying flowering response, fruit development and expression of storage protein genes.

Useful Proteins from Plants

Many plants are now being used to produce useful proteins. This has given birth to

Neutraceuticals – a word coined for made up food. These foods are also known as functional foods. The neutraceuticals include all 'designer' foods from the vitamin-enriched breakfast cereals to Benecol, a margarine spread that actually lowers LDL cholesterol.

Vaccine Production from Plants

Plants are a rich source of antigens for the immunization of animals. Transgenic plants may be developed to produce antigenic proteins or other molecules. Production of the antigen in an edible part of the plant could prove to be an easy and effective mode of delivery system for the antigen in an edible part of the plant could prove to be an easy and effective mode of delivery system for the antigen.

Potential applications of this technology would include efficient immunisation of humans and animals against disease and control of animal pests. For instance, antigens for the Hepatitis B virus have been successfully expressed in tobacco plants and used to immunise mice. Mice fed potatoes expressing the P-sub unit of E.coli enterotoxin LT-B have also produced antibodies, thus protecting against the bacterial toxin.

This technique promises to pave the way for inexpensive immunisation against several human diseases. Oral vaccines against cholera have already been expressed in plants. Generation of antigens through plants is not only cost-effective, but can also be mass produced, and easily recovered.

Non-Transgenic Applications

Marker-Assisted Selection

Plant breeding has always relied on the selection of agronomically favourable characters from the diverse gene pool that is present in any crop species, even if many elite commercial cultivars tend to be based on a very limited gene pool. Often these agronomic characters are visible and easily identified, e.g. height or flower colour or resistance to fungal attack. In other cases, the characters can be much more subtle and sometimes can only be measured by sophisticated analytical techniques, e.g. the amounts of certain secondary products or the fatty acid composition of the seed oil. In all of these cases, it was formerly necessary for the breeder to grow and analyse each new generation before it was possible to measure the character, or phenotype, and select the appropriate plants. The advent of marker-assisted selection has changed this because breeders can now select a few plants that are likely to express the required characters from amongst tens of thousands of progeny even before the plants have developed to maturity. The basis of the method is DNA-fingerprinting and is in principle not different from the methods used to great effect in medical diagnostics or in forensic science.

Molecular markers such as microsatellites, restriction fragment length polymorphisms

(RFLPs) and random amplified polymorphic DNAs (RAPDs) have now been developed for many crops, including trees. These markers can be assembled into genetic maps that have considerable utility, both in basic biological research and in commercial breeding programmes. The markers can be used to track the presence of valuable characters in large segregating populations as part of a crop-breeding programme. For example, if a useful trait such as disease resistance or high oil yield can be linked with a specific marker, many hundreds or even thousands of young plantlets can be screened for the likely presence of the trait without the necessity of growing all the plants to maturity, or doing costly and time-consuming physiological or biochemical assays. Although the earlier molecular markers such as RFLPs were relatively expensive, newer markers such as microsatellites and single nucleotide polymorphisms (SNPs) are considerably cheaper and easier to use. The use of molecular markers can decrease the time scale of crop-breeding programmes by several years and reduce costs. Although largely limited to the major temperate crops at present, the same technology can be applied to assist the breeding of any crop and even to domesticating entirely new crops. A good example of the potential for marker-assisted selection can be seen with tree crops, many of which are major export earners for developing countries (examples include oil palm, coconut, coffee, tea, cocoa and many fruit trees, such as bananas and mangoes). Due to the relatively long life cycles and the formidable bulk of tree crops, research into their biology has always been a relatively difficult undertaking. Hence the application of modern scientific breeding methods, as used for many decades with non-tree crops, is only just beginning to be done for trees. Indeed, despite the tendency to focus on transgenic crops, marker-assisted selection may be an equally significant, albeit less visible and dramatic, application of biotechnology for crop improvement in agriculture.

Tissue Culture and Mass Propagation

The application of cell, tissue and organ culture is central to many modern crop-improvement programmes. Indeed, the limiting step to the successful transformation of most of the major crops has not been transgene insertion itself, but rather the regeneration of viable plants from the transgenic explant material. Tissue culture has been widely used in breeding programmes for over 50 years. For example, the use of embryo-rescue techniques has enabled the introgression of characters such as disease resistance from wild relatives of crops into elite breeding lines. It is now possible to make wide crosses between hexaploid wheat and barley, rye or diploid wheat. The hybrids of such crosses are sometimes sterile due to embryo abortion but can be 'rescued' by culturing or transplanting the embryos.

Another important technique that is increasingly used in crop breeding programmes is the production of doubled haploids. The repeated selection of heterozygous materials in a breeding programme can increase uniformity, but many generations are required to reach homozygosity in loci associated with agronomic traits. The artificial production of haploid plants followed by chromosome doubling offers the quickest method for developing homozygous breeding lines from heterozygous parental genotypes in a single generation. Haploid gamete cells from anther or ovary can be diploidized after colchicine treatment, and then regenerated to yield doubled haploid plants. This technique is now widely used for the improvement of many of our most important crops, including all of the major cereals, potatoes, brassicas and even some trees.

Yet another useful application of tissue culture methods is the mass clonal propagation of certain crops, in particular, trees. The need for the rapid multiplication of millions of seedlings for new biomass crops such as Miscanthus, or other crops such as chicory, has now led to the development of automated methods for their clonal propagation. Clonal propagation has not always been commercially successful, however. In the 1980s, a scheme to mass propagate millions of oil palm plantlets from a superior breeding line foundered when many of the maturing trees were discovered to have an abnormality in their floral development. This led to a failure of fruit formation and, since the major products of the crop are fruit oils, the trees were effectively useless. This abnormality is now known to be due to a tissue culture effect whereby the expression of a homeotic gene regulating meristem identity is disrupted. Although the problem has now largely been rectified, industrial confidence in clonal propagation has not recovered and very little planting of clonal oil palm has been done over the past 20 years.

The continuing scope for crop improvement, following the identification of higher-yielding germplasm, followed by mass propagation can be exemplified once again by considering the case of oil palm. Over the past decade the average yield of Malaysian palm oil has stagnated at 3.5–3.8 ton/ha, although new clonal lines are available that can yield as much as 7.5 ton/ha. Malaysia currently produces about 13.5 million tons per year of palm oil worth an annual $4 billion: this is in a country with a total GNP (gross national product) of $60 billion. To put this figure of $4 billion into context, the estimated entire revenue generated by the US agbiotech sector in 1999 was just $2.3 billion, and this included all the companies supplying inputs to the sector or its employees. Therefore, the effective doubling of the palm oil yield that could be implemented following a successful mass-propagation programme could contribute a hefty 6.6% extra to the national GNP of this one country. The application of a similar strategy with other tree crops, or even relatively undomesticated annual crops, could also yield equally striking results that would particularly benefit developing countries.

Genomics

Genomics is the term given to the massively parallel study of the DNA and protein sequences in an organism, and also when and where such sequences are expressed. However, genomics is much more than the mere assembly of DNA or protein sequence information or gene expression catalogues. It can also be used as a tool in crop breeding programmes and even for the domestication of new plant species as future crops. Many traits of agricultural importance in crop plants appear to be regulated by a large number of genes, and therefore do not segregate into simple Mendelian ratios, as would be expected if only one or two genes were involved. Examples of such complex traits include height, branching, seed oil, protein yield and flowering time. During the past few years, however, the use of more sophisticated genomic tools has shown us that, although dozens of genes may underlie such complex traits, sometimes much of the variation in their phenotypic expression can be caused by a small number of key regulatory genes. These genes can now be identified and mapped based on sequence similarities, expression profiles and molecular markers.

Genes that play a major role in regulating agronomically relevant complex traits in model plants, such as Arabidopsis, and also some crops such as maize, are now being isolated at an ever-increasing pace. Examples include height, flowering time, vernalisation, shattering of seed pods and stem branching. Many of these plant genes encode transcription factor proteins that in turn regulate the expression of large sets of other genes. For example, transcription factors can switch on entire metabolic pathways or patterns of cell division, resulting in the formation of new tissues or organs and the accumulation of new storage products.

The isolation and mapping of important regulatory genes from a model plant species such as Arabidopsis or rice will soon enable the equivalent gene, in terms of both sequence and chromosomal location, to be isolated from other plants, including most of the major crop species. This is due to a striking feature of many higher plant genomes, namely their high degree of similarity to one another in terms of the structure of their genes and, in related species, the order in which they occur on chromosomes. This gene order, or synteny, is particularly well conserved in monocotyledonous plants, which include all of the cereal crops. Recent evidence suggests that there is also appreciable genome synteny within the dicotyledons, which include the important model plant Arabidopsis as well as major crops such as soybean, rapeseed and tomato.

The long-term implications of these developments are profound. They open up the prospect of being able to manipulate some of the most basic features of crop plants. The application of genomics for crop improvement need not involve transgenesis. Key agronomic traits identified from model plant genomic studies can often be mapped to their syntenic equivalents in a crop of interest. This information could then be used to generate molecular markers to assist breeding programmes seeking to manipulate such traits. In annual crops, the height and branching could be adjusted to optimise harvestability and the capacity of the plant to bear and retain its seed. The seeds could be selected to accumulate the exact mixture of carbohydrate, protein, oil and fibre required for a particular end use. Flowering time could be adjusted to suit particular climatic conditions. For example, the development of earlier flowering varieties of some crops in a location such as the Canadian prairies would allow them to set seed and be harvested before the onset of early frosts that can otherwise halt seed development before it is completed. A slight advancement of the flowering time could also dramatically improve yields of rice in some tropical and subtropical regions where the current growing season is just over six months. If the length of the growing season could be reduced to less than 6 months, the farmers in such regions could grow two rice crops in each year. The development of shorter varieties of cereal grain crops in the 1960s and 1970s led to such huge increases in yields that it was dubbed the 'Green Revolution'. However, some crop varieties, including the highly prized Basmati rice, have remained recalcitrant to efforts at introducing dwarf traits by conventional breeding. The ability to alter height by gene manipulation is therefore an attractive prospect.

Transgenic Applications

The Basic Technology

The widely used Agrobacterium vectors for the delivery of transgenes into plants were developed almost two decades ago. It is therefore surprising that this method of gene insertion has undergone relatively little subsequent refinement. Transformation of plants is still fairly crude and inefficient compared with, for example, the transformation of microbes or animals. Transgenes in plants insert more or less randomly into the genome, the number of copies inserted can range from one to a dozen or more, and the introduced DNA sequences are frequently modified and/or fragmented in the process. For example, pieces of DNA may lose part of one or both of their end regions, or they may be cleaved at several points and the resulting fragments inserted into different parts of the genome. Both the position of insertion in the genome and the copy number of a transgene can significantly affect its expression in the resulting plant. The presence of multiple copies of a transgene can result in instability of its expression. Finally, transgenes are inserted into the recipient genome as part of a multigene construct that also contains regulatory elements and a selectable marker, often an antibiotic- or herbicide-resistant gene. There is already public concern about the use of antibiotic-resistant markers, and many researchers and breeders are also concerned that, as transgenic crops become more widespread, they will inevitably subject to many further rounds of transformation as additional genes are inserted to keep improving the crop. The accumulation of many different selectable marker genes would soon present problems as breeders run out of new benign markers for further rounds of transformation. The continued use of the same marker can lead to co-suppression and the loss of expression of the transgene itself, and also of genes with related sequences.

These drawbacks to existing transformation technology highlight what should be a major research priority in the future. It is striking that compared to the enormous investment in all the new 'omic' technologies (genomics, proteomics, metabolomics, etc), much less attention has been paid to the possibly less glamorous, but certainly fundamentally important, technologies involved in gene insertion into plants. A key priority should be the development of a method in plants for the facile site-specific insertion of genes, as already exists for animals. Extraneous DNA, e.g. antibiotic-resistant genes or other markers, should be removed from constructs after they have been inserted into the plant genome. Several methods, such as the cre-lox system, have been already developed for this purpose, but these need to be refined and made easier to use. Another potentially promising development is the FLP-recombinase system of yeast. Several new selectable marker systems, such as threonine dehydratase/deaminase from Arabidopsis thaliana, or phospho-mannose isomerase are also being developed. An alternative to selectable markers is the use of 'scoreable' markers, which encode enzymes not normally present in the plant and whose activity can easily be measured. One example is the development by Pioneer (now owned by DuPont) of a transgenic sunflower variety where the marker was oxalate oxidase, which can be detected by a simple and sensitive assay.

The development of relatively few commercial transgenic crop species to date is partially explained by the technical difficulties (until recently) in efficiently making transgenic forms of a range of varieties of some of the other major crops, such as wheat and barley. This was compounded by an understandable focus in the 1990s on the development of mainstream profitable crops by commercial companies, rather than less profitable staples. In the past few years, the use of new strains of Agrobacterium plus developments in tissue culture has resulted in the successful transformation of the major cereals, such as wheat and rice, by this method. This promises to facilitate future development of transgenic cereal crops. Although the transformation of virtually all of the major annual crop plants has now been achieved, in many cases this remains a time-consuming and costly technical process. For example, the transformation of soybean by gene transfer into cultured embryos is a relatively lengthy, inefficient and labour-intensive process. Even in some of the species, such as rapeseed, where transformation is relatively facile, it is often highly cultivar dependent. This means that it may not be feasible to transform the latest elite cultivars of rapeseed with a gene or genes of interest. Instead, older varieties such as Westar, which has already been optimised for transformability, are often used. To produce a commercial transgenic variety, the Westar transformant must then be backcrossed to an elite line—a process that can take several years and add significantly to development costs. By the year 2000, more than 25 tree species had been transformed with various genes, most of which were marker (40%) or herbicide-tolerant (24%) genes. Although there have been over 150 field trials of transgenic trees, none of these has yet resulted in extensive commercial planting.

Current Applications of Transgenic Crops

The first transgenic crop to be commercially cultivated was the (now defunct) FLAVR SAVR™ tomato that was originally approved for cultivation by the USDA in 1992, and finally authorised for human consumption by the FDA in 1994. It is noteworthy that, almost a decade later, over 99% of all the transgenic crops in the world are being grown in just four countries (68% in the USA alone). Furthermore, virtually the entire global acreage of transgenic crops in 2001 consisted of only four species, namely soybean, maize, cotton and rapeseed (International Service for the Acquisition of Agri-biotech Applications Server. Finally, only two classes of traits have been manipulated in all of these transgenic crops, i.e. herbicide tolerance and insect resistance. Hence, although the global acreage of transgenic crops has shown a steady increase to the impressive figure of 52.6 Mha in 2001, the range of crops, traits and growing areas remains remarkably narrow. The complete data for transgenic crop cultivation in 2001 are shown in figure.

All of the current first-generation transgenic crops, i.e. those in commercial production from 1996 to 2002, have been engineered for altered input traits. Input traits are those characteristics of a crop that affect its cultivation and yield, but not the quality of the products from the crop. Such traits are typically related to agronomic inputs, such as herbicides, pesticides and fungicides, as well as nitrate and mineral fertilizers. Input traits were the first commercial target of transgenic manipulation in crops for several reasons. First, the traits can often be modified by the insertion of a single gene and many of the relevant genes had already been isolated by the mid-1980s. Secondly, seeds harbouring the new traits would have an added value that could be readily captured by the company that developed them, hence quickly offsetting the R&D costs and then generating a continuous net revenue stream. Thirdly, the result of the expression of such traits should be better yields and/or lower production costs; no changes in crop-harvesting methods are needed and no new crop products have to be processed. This made the development of transgenic crops with enhanced input traits an attractive short-term proposition for the seed companies which developed these first-generation genetically manipulated (GM) crops. The main commercial downside of this strategy has been that consumers in industrialised countries do not perceive a direct benefit to themselves from such crops, which produce the same foods at the same price as conventional crops. At the same time, many consumers (especially outside the Americas) have perceived a risk from GM-derived foodstuffs. This has now delayed the adoption of the transgenic crops in Europe, possibly for several years to come. However, one of the factors behind the increasing acceptance of transgenic crops in Asia appears to be the direct benefits to low-income farmers in terms of reduced input costs and higher yields. Hence, the dichotomy between the acceptance of GM crops (in the United States and elsewhere) and their non-acceptance (in Europe) may be at least partially explained by more effective producer lobbies in the former countries compared with more effective consumer lobbies in Europe.

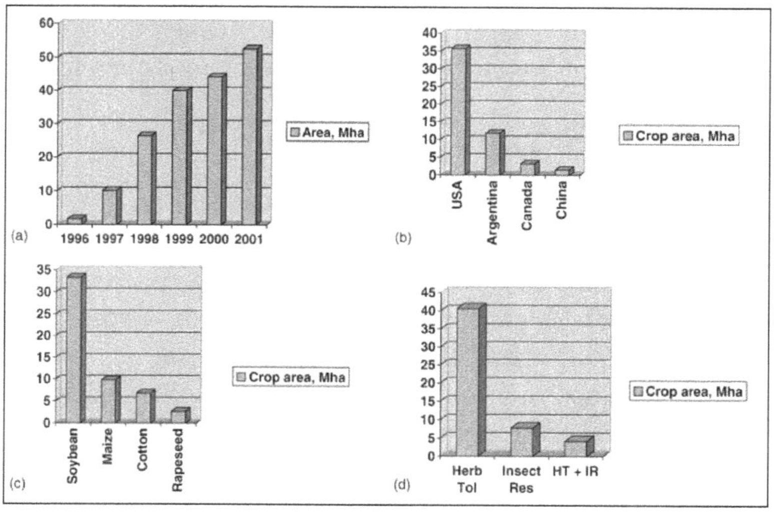

Figure: (a) Global area of transgenic crop cultivation, 1996–2001, (b) cultivation areas of transgenic crops in 2001, by country, (c) cultivation areas of transgenic crops in 2001, by crop species, (d) cultivation areas of transgenic crops in 2001, by trait.

Herbicide Tolerance

Weeds are plants that compete with the crop of interest, reducing its yield and complicating harvesting. Crop plants themselves can be serious weeds, e.g. volunteer cereals in rapeseed fields and vice versa. Weeds are conventionally controlled by the serial application of herbicides that may be more or less specific, i.e. they may damage the crop to some extent as well as eliminate the weeds. The development of crops that are resistant to some of the most powerful broad-range herbicides was one of the earliest targets of transgenic research. Of the total global output of 52.6 Mha of GM crops in 2001, over 77% (by area) were modified for herbicide tolerance alone while a further 8% were modified for both herbicide tolerance and insect resistance.

The most widespread types of transgenic herbicide-tolerant crops are those developed by Monsanto under the trade name of 'Roundup Ready'. Such crops are resistant to the widely used broad-spectrum (i.e., non-specific) herbicide, glyphosate, which is marketed by Monsanto as 'Roundup'. Glyphosate is a toxin that inhibits the enzyme 5-enolpyruvylshikimate-3-phosphate synthetase (EPSPS) in plants, resulting in a lethal disruption in their ability to synthesise proteins. Although all known plant versions of EPSPS are highly sensitive to inhibition by glyphosate, many bacteria have a slightly different form of the enzyme that is insensitive to the herbicide. Therefore, if a copy of a bacterial EPSPS gene is inserted into a crop plant, the resulting transgenic crop variety will be resistant to applications of glyphosate while all non-transgenic plants in the same area will be killed. The cultivation of transgenic glyphosate-resistant crops is claimed to result in significant financial benefits for farmers because of reduced overall herbicide applications and higher yields per hectare (reportedly worth $15–28/ha). For their part, the seed companies are able to sell a profitable package, including seeds and herbicide, to the growers.

A similar package of herbicide-resistant transgenic crops, plus the related herbicide, has been developed by AgrEvo under the trade name of 'Liberty Link'. In this case, the herbicide is the fungal toxin, glufosinate, which is marketed under several names including 'Basta' and 'Challenge'. This widely used compound is an inhibitor of glutamine synthase in plants, causing a disruption of photosynthesis that results in the death of the entire plant within a few days. Glufosinate is a broad-spectrum herbicide, and its use is therefore limited to total eradication of vegetation or to control of weeds shortly after crop emergence. In contrast, glufosinate-resistant crops can be sprayed with the herbicide at any time, resulting in the effective elimination of all other plants from the field. Resistance to glufosinate is conferred by the addition of a gene from the soil-dwelling bacterium, Streptomyces viridochromogenes, which encodes the enzyme phosphinothricin acetyltransferase. This enzyme is able to acetylate glufosinate, which results in the loss of its toxic activity. Plants expressing this transgene are therefore able to grow normally, even after the application of relatively large doses of glufosinate. Transgenic glufosinate-resistant rapeseed was first grown commercially in Canada in 1995, soybean and maize were approved in 1997 and other crops such as sugar beet will soon be commercially available.

There are concerns that in some crops such as rapeseed or sugar beet, which have closely related weed species, the herbicide-resistant trait may spread into the weed population by cross-pollination. However, non-transgenic herbicide tolerance has also been developed in rapeseed and other crops. For example, the so-called SMART-canola varieties of the Canadian rapeseed developed by American Cyanamid (now owned by BASF) are resistant to imidazolinone herbicides but are the products of conventional (i.e., non-transgenic) plant breeding. These crops have been grown since 1995 and their use is widespread in Canada (Biotechnology Industry Organization Server. In principle, such non-transgenic herbicide-resistant crop varieties are as likely to outcross and transfer their resistance traits to weedy relatives as the transgenic varieties. Therefore, concerns about the spread of traits such as herbicide tolerance should not only be related to the use of transgenic crops per se, but rather should involve more general management of any agronomic trait where its transfer to weedy relatives, or elsewhere, may have deleterious consequences. This is an important issue that often seems to escape the attention of environmentalists and government regulators alike.

Pest and Disease Resistance

Crop yields are regularly reduced by herbivore attack, most notably from insects, and by diseases caused by nematodes, fungi, bacteria or viruses. Even in intensively managed agronomic systems, pests and diseases can routinely cause losses of >30% of total yield potential. Meanwhile in developing countries the figures are often much higher. After herbicide tolerance, the second most common modification in transgenic plants in 2001 was insect resistance, which alone accounted for 7.8 Mha, plus a further 4.2 Mha where the trait was expressed in combination with herbicide tolerance. Insect resistance in transgenic maize, cotton and potatoes was conferred by insertion of a gene encoding a protein toxin from the Gram-positive soil bacterium Bacillus thuringensis (Bt). The use of insect-control sprays containing a pro-toxin-producing Bacillus thuringensis suspension has been common for over 30 years in organic farming, but the widespread use of Bt toxins in transgenic crops is much more recent. The Bt toxins are a family of so-called crystalline (cry) proteins that are converted into their active form during digestion in the gut of a range of insect larvae, resulting in a disruption of potassium ion transport that rapidly becomes lethal. Mammals do not convert the pro-toxins into their active forms and are therefore unaffected by them. Early indications suggest that transgenic Bt crops are effective in controlling insects and improving yields (by 7%), while also reducing the need for spraying with more toxic and less desirable pesticides that often affect beneficial organisms such as insectivorous birds (saving growers some $40/ha).

The obvious danger in relying on a single class of toxins is that it tends to establish a strong selection pressure favouring the survival of insects that are able to sequester the toxin, or otherwise render it harmless. Before the mid-1990s, only a few thousand hectares of land on organic farms were sprayed annually with live suspensions of Bt, but in the past few years the cultivation of transgenic Bt crops has expanded to over 12 Mha and this area is still increasing. The resultant chronic and widespread exposure of hundreds of insect species to the Bt toxin must increase the likelihood that resistance will eventually develop in some species. Indeed, as long ago as 1995, at least two insect species had already become resistant to Bt toxins in the field with at least another ten species showing the potential for the acquisition of resistance in laboratory studies.

Another risk with the expression of xenoproteins, such as the Bt toxin, in crops is that their accumulation may be curtailed if the plants are stressed. For example, in 1996 a combination of heat and drought caused a reduction in levels of Bt toxin in Monsanto's transgenic 'Nu Cotn' variety of cotton. Although deleterious to the crop, these climatic conditions actually favoured the development of the cotton bollworm, resulting in severe infestation in nearly 1 Mha of 'resistant' cotton crops. The infestation was eventually controlled with conventional chemical pesticides, but the more serious long-term danger is that exposure of the bollworms to sublethal doses of the Bt toxin would be an ideal mechanism for the development of resistance by the insects.

To combat the development of resistance to Bt toxins in the field, growers are now advised to set aside refugia. The refugia are areas adjacent to the main transgenic crops that are sown with non-transgenic seeds of the same crop. In these areas, non-tolerant insect populations can continue to thrive and hopefully will out-compete those of their conspecifics that develop Bt tolerance. This strategy relies upon the cooperation and enforcement of good management practices by growers, and can fail if it is not implemented rigorously. Some of the recent experiences of breakdowns in communications between seed companies and farmers growing transgenic crops (e.g. the STARLink maize affair of 2000–2001) serve to underline the potential difficulties in maintaining an effective refugia policy. Another strategy to prevent or delay the acquisition of resistance by insects is the inclusion of several unrelated toxin genes in a transgenic crop, but this so-called gene pyramiding will be a more expensive and long-term option. An alternative strategy to delay the onset of Bt-resistant insect populations in crops is to sow mixtures of Bt and non-Bt seeds so as to provide the opportunity for some non-resistant insects to survive and compete with resistant insects. Recent studies have found that, while it did delay the acquisition of Bt resistance by the European corn borer, a seed mixture strategy was less effective than the refugia method.

Viruses, bacteria, fungi and nematodes are major pathogens of crops and there has been a great deal of research aimed at producing resistant varieties by transgenic approaches. While various fungicides and nematicides are available to help farmers control these pathogens, there are no equivalent virus-control agents, so the combating of viral diseases normally relies on the endogenous resistance of the plant. In the absence of endogenous resistance, viral infections can be particularly devastating to a crop. This has stimulated efforts to engineer viral resistance into transgenic crops.

The commercial cultivation of several transgenic potato, squash and papaya varieties with virus-resistant genes has already been approved in some countries. These transgenic crops express complete or partial proteins from a particular virus (typically part of the viral coat protein complex), which causes the plants to become sensitised to subsequent infections with the same virus. When such plants are attacked by the viral pathogen, they mount a successful defence response in a manner that is somewhat analogous to immunisation in animals that have been injected with an attenuated virus, although the exact mechanism of viral immunity in plants remains to be explained. A recent example of a transgenic virus-resistant crop is a variety of papaya developed in Hawaii and Australia. The papaya ringspot virus is a major threat to the cultivation of papayas in tropical countries such as Hawaii. Transgenic papayas that express the ringspot virus coat protein, which on its own is harmless to the plants, are considerably more resistant to infection with the active virus than are non-transgenic papayas.

The engineering of resistance to bacterial, fungal and nematode pathogens has been more problematic although several promising approaches have been demonstrated, at least in principle. One possibility is the expression of resistance genes such as the Xa21 bacterial blight-resistant gene that was recently transferred to five Chinese rice varieties. Anti-fungal agents such as phytoalexins or chitinases have also been expressed in plants. In another case, a gene encoding an anti-fungal protein from alfalfa was transferred to potatoes. The transgenic potatoes were resistant to the soil-borne fungus Verticillium dahliae, which causes some $70–140 million in damage to potato crops each year. However, this resistance was quite specific and the potatoes were still susceptible to the serious pathogen, Phytophthora infestans, the late-blight fungus that caused the devastating Irish famine in the 1840s. This highlights one of the difficulties with engineering fungal resistance in crops, namely that it is very difficult to produce broad-spectrum resistance, and so to achieve this it may be necessary to transfer numerous resistance genes. Also, fungal resistance often evolves naturally and can be found in different varieties of a crop or in sexually compatible wild relatives, from which it can be transferred to an elite crop cultivar by conventional breeding. In the above case, a potato variety called Russet Burbank was transformed with an alfalfa gene so as to confer resistance to the fungus, Verticillium dahliae, but there is another potato cultivar called Russet Ranger that is naturally resistant to this fungus anyway.

Nematodes are the major class of animal parasites of crops, causing over $100 billion in annual losses to world agriculture. Natural resistance to nematodes is relatively restricted, while chemical control agents such as carbamates include some of the most toxic pesticides in widespread use, and are both costly and environmentally damaging. The development of transgenic crops with nematode resistance could, therefore, have appreciable economic and environmental benefits. Some of the research approaches include the induction of 'suicide genes' in plant cells infected with a nematode, or the expression in plants of protease inhibitors that inhibit nematode growth.

Other Input Traits

Numerous additional input traits of interest to seed companies are being developed using transgenic plants. Male sterility is desirable for the production of hybrid crops, which normally have considerably higher yields compared with their non-hybrid counterparts. Male sterility can also have an environmental benefit since the pollen will be either absent or sterile. This can be a useful trait in transgenic crops, as it will prevent cross-pollination and the spread of the transgenes into other related species.

A well-established way of producing male sterility, which has long been used by plant breeders, is to prevent the formation of viable pollen grains during their development in the anther. In the context of transgenic crops, this has been done by inserting a cytotoxic transgene under the control of an anther-specific promoter into rapeseed. Expression of the transgene is restricted to the pollen-nourishing tapetal cells of the anther and prevents the formation of mature pollen grains: this renders the male plants sterile. A popular cytotoxic gene of choice is barnase, which is obtained from Bacillus amyloliquefaciens and encodes a ribonuclease that destroys RNA, hence killing any cells in which it is expressed. The ribonuclease is unable to enter adjacent cells so that its lethal effects are limited to those cells in which the barnase gene is expressed. In a further refinement of this approach, the addition of a second gene called barstar causes the barnase toxin to be disabled, which allows the pollen grains to develop, hence restoring fertility. This provides the breeders with the male-sterile and male-fertile genotypes required for the production of commercial hybrids. Although this technology has been available for nearly a decade, it is still undergoing field trials in various countries although there are good prospects that it will be commercialised soon.

Another well-known transgenic sterility trait is that conferred by the unfortunately named 'terminator' technology which was being developed by several companies, including the one acquired by Monsanto for commercial release in the late 1990s. However, Monsanto withdrew the technology (temporarily) from use in crops late in 1999 following adverse public reaction. This technology is a series of approaches that have numerous possible manifestations (as well as possible drawbacks). In one of the applications, a gene would be inserted into a crop such that, as the seeds germinated or the seeds from the second generation developed, a toxic protein would be produced, and the seedlings or the seeds would die. Its main benefit would be to seed companies since farmers would not be able to save seed for replanting in subsequent years (as they have traditionally done in most developing countries), but would instead have to re-purchase the seed each season from the company. An additional more general benefit would be to prevent the spread into the environment of transgenic traits such as herbicide tolerance that may be present in such seeds. Several major agbiotech companies are continuing to research sterile-seed technologies, and in August 2001 the USDA licensed a similar technology to Delta &

Pine Land, albeit with the condition that it should not be used commercially before 2003.

Output Traits

The manipulation of output traits is often considerably more complex than that of most input traits. Whereas most input traits can be usefully expressed constitutively, i.e. throughout the plant, new output traits are normally desired only in the harvested portion of the crop, e.g. the seed or fruit, and may even be deleterious if expressed elsewhere. Output or quality traits are often the products of complex metabolism and may require the insertion of several transgenes to have an effect. Finally, and most importantly, the modification of an output trait will, by definition, result in a new crop variety with different products to non-modified varieties. Unless the transgenic crop completely replaces non-transgenic varieties, it will require complete segregation at every stage of production from seed storage and planting to harvesting and downstream processing. This can add at least 10–20% to costs and imposes considerable (and often overlooked or underestimated) management problems. The predominant agriculturally related output traits that are currently under development relate to the major seed and fruit storage products, i.e. starches, oils and proteins.

Storage/Shelf Life

Many losses of crops occur because of spoilage during storage or transportation. Often such losses can be reduced by harvesting relatively unripe crops, especially fruits, and ripening them when required by applying agents such as the plant hormone ethylene. A drawback of this process is that the unripe fruits have often not developed their full range of flavours and these are not always induced by artificial ripening. One has only to taste the thick-walled, long shelf-life tomatoes from supermarkets, and compare them with shorter lasting but much more highly flavoured homegrown produce. The main reason for the deterioration of soft fruits during storage and transport is a decline in firmness, which increases the chances of injury and infection of the fruits. The fruits become softer by cell wall breakdown and one of the key enzymes in this process is polygalacturonase.

One approach to slowing down the softening of fruits is to decrease the activity of polygalacturonase during ripening. This could allow fruits such as tomatoes to remain for longer on the vine in order to develop more flavour, but without the risk that they would lose the firmness necessary for them to be harvested, transported and stored without damage. The isolation of the polygalacturonase gene in the 1980s allowed this approach to be adopted by Calgene in the USA, and Zeneca in the UK. In both the cases, an antisense (i.e. back-to-front) copy of the gene was inserted into tomatoes. The result was the first food crop to be produced by transgenic manipulation, i.e. the FLAVR SAVR™ tomato, which was released for human consumption by Calgene, in 1994. The

FLAVR SAVR™ tomatoes were claimed to taste better and last longer in storage than conventional tomatoes, and initially sold well in supermarkets in the USA. Ironically, FLAVR SAVR™ tomatoes were not a success for rather mundane reasons that were unrelated to GM technology but are a useful lesson for future attempts to commercialise such crops.

In 1995, the UK-based Company Zeneca released a related transgenic food product. This was a tomato paste from tomatoes in which the enzyme polygalacturonase had also been down-regulated so that the paste was claimed to be appreciably thicker and better tasting. Unlike the FLAVR SAVR™ fresh tomatoes, the Zeneca tomato paste was a modest commercial success, capturing over half of the market where it was sold, until it was overwhelmed in the anti-GM sentiment that swept the UK early in 1999. The tomato paste was withdrawn from sale by supermarkets even before the public registered any disapproval of the product. Unlike the current generation of crops with transgenic input traits, these two examples of transgenic tomatoes with modified output traits were segregated from non-transgenic foodstuffs and were distinctively labelled as GM products.

Future Applications of Transgenic Crops

As we progress beyond the first-generation portfolio of transgenic crops, it may be useful to take stock of the potential for the development of new types of transgenic crops over the next decade or so. Emerging developments in the application of genomics to agbiotech may allow for some radical modifications of plant architecture, growth habit and composition. Another important group of traits that is under particular scrutiny is that relating to abiotic stress.

Abiotic stresses seriously affect crop performance in the field almost everywhere. If short-term (i.e. within a few decades) global climate change is indeed a reality (which remains to be demonstrated conclusively), the incidence of all forms of biotic and abiotic stresses in agricultural systems is likely to increase significantly. Even if there are no major climatic changes over the next 50 years, environmental stresses are increasingly affecting agriculture. In some regions, changing patterns of land and water use are causing increased salinisation of agricultural areas, while elsewhere there are serious problems of mineral toxicity. Drought and salinisation are already the most common natural causes of famine in arid and semi-arid regions, and are the most significant threats to agriculture in many parts of the world. It has been estimated that as much as 30% of arable land may be lost to salinisation over the next 25 years. The resulting conflicts over water resources at both local and international levels could exacerbate food shortages still further in the affected regions.

Drought tolerance arises from a complex set of traits that may have evolved as separate mechanisms in different plants. In view of the likelihood of more arid regions arising in the near future, it is surprising that there have been relatively few attempts to produce transgenic drought-tolerant crops, even by publicly funded organisations. An alternative approach to transgenesis is to use advanced breeding methods to improve the agronomic performance of existing arid-region crops, such as pearl millet, which is grown on over 40 Mha in Africa. The synteny of the pearl millet genome with other major cereals gives some hope that drought-tolerant traits will eventually be introduced into local varieties by conventional breeding with the help of some of the newer and less expensive molecular marker maps.

Several studies are beginning to show the potential of a transgenic approach to the challenge of developing tolerance to abiotic stresses in crops. One example is a transgenic tobacco line, expressing an E. coli mannitol-1-phosphate dehydrogenase gene, that accumulates elevated levels of mannitol and is therefore better able to withstand high salinity Laboratory and small-scale field studies have shown that the accumulation of other compounds, including betaine or trehalose, in transgenic plants may also enhance salt tolerance. More recently, a rapeseed variety expressing an Arabidopsis vacuolar transport protein was shown to tolerate as much as 250 mM NaCl without significant impact on its seed yield or composition. However, it is not clear whether such relatively simple modifications will lead to a sustained effect on crop yields in the much more complex real-life cropping systems where osmotic stress is often linked with a combination of other factors, such as periodic aridity, mineral/salt build up and/or erosion.

Some plants are able to withstand relatively high levels of mineral toxins, such as heavy metals. The metals are often absorbed by these plants but are then chelated to specific proteins or other compounds, which allows them to be sequestered in a non-toxic form. The ability to express such traits in crop plants could extend their range of cultivation, or may allow them to be used for reclaiming polluted land. Another intriguing idea is to use such plants for 'bioprospecting'. This could involve the cultivation of a metal-sequestering crop in an area that contains valuable mineral deposits that are not sufficiently abundant to justify direct mining. The possibilities of such an approach are illustrated by a report that a brassica species has been used to hyper-accumulate gold. One of the best-studied classes of proteins associated with heavy-metal tolerance is the metallothioneins, which can bind metals such as Zn, Cd and Cu. Transgenic rapeseed or tobacco plants expressing mammalian metallothioneins can tolerate elevated levels of heavy metals. There have also been efforts to overexpress another class of glutathione-derived metallothioneins, termed phytochelatins, by transferring a bacterial γ-glutamylcysteine synthase gene into poplar trees. Preliminary data indicate that the transgenic trees contain higher levels of glutathiones, but it is not yet clear whether this leads to increased phytochelatin accumulation or tolerance to heavy metals.

Although these approaches are yielding promising early results in lab studies, it will be necessary in future to carry out a thorough analysis of the field performance of such transgenic plants in order to assess possible pleiotropic effects. This is particularly relevant to the over-expression of metallothioneins because the precise biological function(s) of these proteins is still unclear, and they may be involved in numerous processes other than metal tolerance. Recent studies have shown that metallothioneins are highly expressed during fruit ripening, leaf senescence, wounding and viral infection and fungal and bacterial infections. Indeed, it is quite possible that the primary role of metallothioneins in plants is related to a process, such as redox regulation rather than metal sequestration. This is a useful example of how our emerging, but still imperfect, knowledge of many aspects of plant physiology and biochemistry is revealing that some of the assumptions that have informed strategies for plant manipulation by transgenesis may require some revision.

Agricultural Biotechnology

Agricultural biotechnology is a set of tools and disciplines meant to modify organisms for a particular purpose. That purpose can include anything from coaxing greater yields from food crops to building in a natural resistance to certain diseases. Though there are multiple ways to accomplish this goal, the method that tends to get the most attention from the public is genetic modification. It is the use of different scientific techniques to modify plants and animals. The undesirable characteristics like susceptibility to diseases and low productivity are bred out. If there is a particular trait that the plant or animal can benefit from, it can be bred in by using a gene that contains the characteristic.

Agricultural biotechnology has been practiced for a long time, as people have sought to improve agriculturally important organisms by selection and breeding. Today, this technology has reached a stage where transgenic plants have been developed as a result from genetic engineering experiments in which genetic material is moved from one organism to another, so that the latter will exhibit a desired characteristic. The common plant transformation techniques are as follows:

Agrobacterium Mediated Gene Transfer

This method of transformation is the most widely used to introduce foreign genes into plant cells. Agrobacterium tumefaciens is capable to transfer a particular DNA segment (T-DNA) of the tumor-inducing (Ti) plasmid into the nucleus of infected cells where it is subsequently stable integrated into the host genome and transcribed, causing the crown gall disease. The T-DNA contains two types of genes: the oncogenic genes, encoding for enzymes involved in the synthesis of auxins and cytokines and responsible for tumor formation; and the genes encoding for the synthesis of opines. Outside the T-DNA, are located the genes for the opine catabolism, By removing the tumor inducing genes and replacing them with the genes of interest, transformed plasmid can be obtained. Thus, making the bacteria an excellent vector for the transfer of foreign DNA.

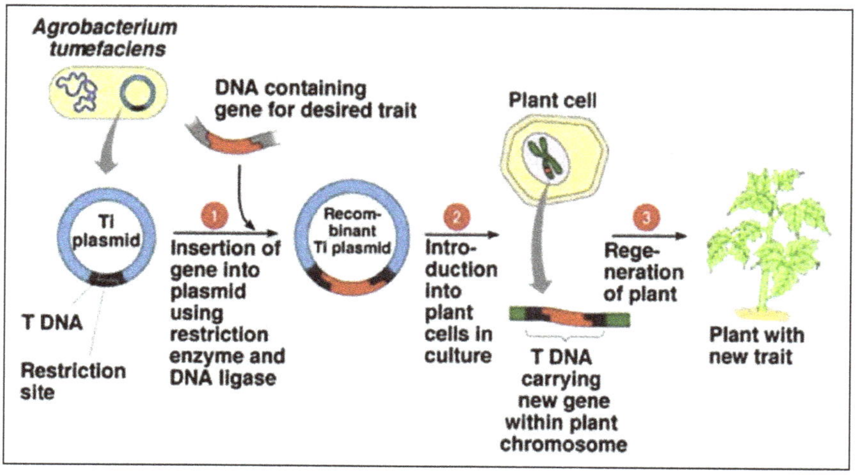

Biolistics

Some cells, tissues and intracellular organelles are impermeable to foreign DNA, especially plant cells. To resolve this problem in gene transfer, the gene gun machine is used for the transfer of desired gene. In this method, DNA or RNA adheres to biological inert particles i.e. gold or tungsten. DNA-particle complex is put on the top location of target tissue in a vacuum condition and accelerated by powerful shot to the tissue and then DNA will be effectively introduce into the target cells.

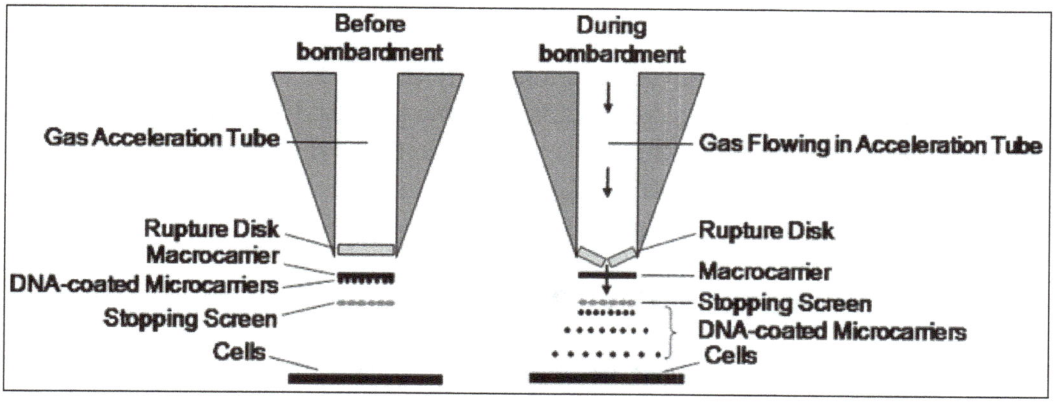

Electroporation

Electroporation is the process where cells are mixed with a DNA construct and then briefly exposed to pulses of high electrical voltage. The cell membrane of the host cell is penetrable thereby allowing foreign DNA to enter the host cell. Some of these cells will incorporate the new DNA and express the desired gene.

Microinjection

Microinjection is the direct mechanical introduction of DNA under microscopical control. A target can be a defined cell within a multicellular structure such as embryo, ovule and meristematic cells or a defined compartment of a cell. By examination with a microscope, a cell is held in place with gentle suction while being manipulated with the use of a blunt capillary. A fine pipet is then used to insert the DNA into the cytoplasm or nucleus. This technique is effective with plant protoplasts and tissues.

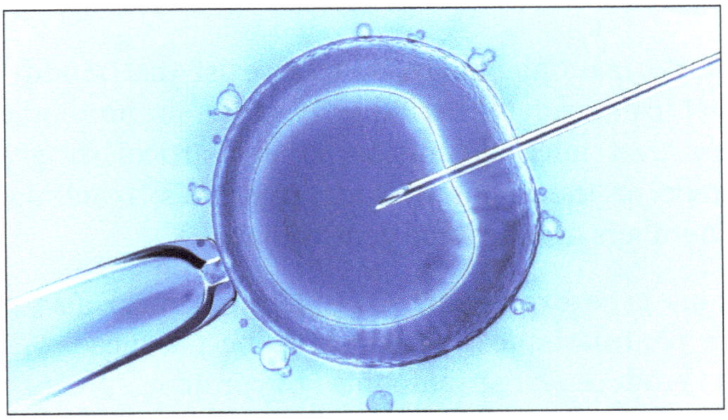

PEG Mediated Gene Transfer

Liposomes are artificial lipid vesicles surrounded by a synthetic membrane of phospholipids, which have been used in animal cell culture for the drug delivery system. This process generally involves three steps:

- Adhesion of liposomes to the protoplast surface.
- Fusion of liposomes at the site of adhesion.
- Release of plasmid inside the cell.

Strategies

Strategies are adopted to obtain the best quality and quantity of the desired product. Following three strategies are:

1. Plant gene expression: Depending upon the purpose for which the product is to be made, required gene expression can be used among the following three methods of gene expression.

2. Transient expression: With transient expression (TE), a gene sequence is inserted into plant cells, without incorporation of the new genetic material into the plant chromo-some. TE systems can be rapidly deployed and can produce large amounts of protein, but because non-chromosomal DNA is not copied with the process of mitosis or meiosis, gene expression is neither permanent nor heritable. While TE

systems are very useful for research and development, and may be useful for drug production, they re-quire the fresh production of transformed plants with each planting and may be less attractive for long-term or high-volume protein production.

3. Stable gene expression: Alternatively, the primary plant chromosome can be altered to allow for the permanent and heritable expression of a particular protein. While permanent modification of the plant genome is more costly and time-consuming, it offers the clear advantage of stable, ongoing protein production with repeated plant-ing alone.

4. Chloroplast transformation: Finally, systems exist that modify chloroplast DNA in plants and that can lead to heritable changes in protein expression. These provide advantages over nuclear transformation, particularly given that each cell may carry hundreds or thousands of such organelles, resulting in the ability to sustain very high numbers of functional gene copies.

5. Location of Trans gene expression: Consideration must also be given to where within the plant a pharmaceutical protein is to be produced. While production in green mass can produce large amounts of protein, green matter is highly physiologically active and protein levels may be poorly preserved if materials are not rapidly dried or otherwise inactivated. Thus, unless a protein or peptide is highly stable, green matter production may result in poor protein recovery and usually requires immediate processing. Tuber or root production, while feasible, shares many of the characteristics of green matter production systems. Unlike green matter, seeds generally contain fewer phenolic compounds and a less complex mixture of proteins and have specifically evolved to provide for stable, long-term storage of proteins and other materials in order to assure success-ful, delayed germination.

6. Selection of plant species and characteristic: It is also necessary to decide which plant species to transform for production of a specific pharmaceutical product. While nearly any plant could theoretically be transformed, practical considerations suggest the use of plants with which we are most familiar, and which already have well-established techniques for genetic transformation, high volume production, harvest, and processing. For green matter production, tobacco has usually been the material of choice, largely because of its highly efficient production of biomass, although other systems such as alfalfa and even duckweed show promise. For seed production, a plant optimized for large seed and high protein production is clearly preferred. Food crop plants have been bred specifically to produce highly productive stands of high-protein seed for which harvesting, processing, and storage technologies are already available.

Transgenic Crops: For Enhanced Quality and Quantity

Various crops have been genetically modified such as to have following beneficial aspects.

Table: Transgenic crops and their transgene for various beneficial aspects.

Crop	Quality character/Controlled pathogen	Gene
Soybean	High oleic acid content	Fad2 (fatty acid desaturase) from Borage officinalis.
Golden rice	Vitamin A	Phytoene synthase (Phy) from daffodil and phytoene desaturase and Zeta-carotene desaturase (Crt1) from Erwinia uredovora and lycopene β-cyclase from daffodil.
Tomato	Increased sucrose level	Maize sucrose phosphate synthase (SPS).
Tobacco	High methionine content	2S albumin of Brazil nut under the promoter of phaseolin gene.
Tomato (Flavr Savr)	Shelf life	Antisence PG (Polygalacturonase).
Tomato (Endless Summer)	Shelf life	S-adenosyl methionine (SAM) hydrolase from T3 bacteriophage.
Tobacco	Tolerance to salinity	Mannitol-1-phosphate dehydrogenase (mt/D) gene from E. coli.
Tobacco	Resistance to drought	SacB from Bacillus subtilis.
Tobacco	Oxidative and chilling stress tolerance	Cu/Zn superoxide dismutase from rice.
Cotton	Against lepidopteran	Bt gene from B. thuringiensis.
Rice	Coleopteran (Sitophilus zeamais)	Corn cysteine
Sugarcane	Sugarcane borer (Diatraea saccharalis)	Bt cry l
Tobacco	Spodoptera litura	Trypsin inhibitor gene (sweet potato).
Potato	Phytophthora infestans	1,3-β glucanase
Tobacco	Alternaria longipes	Bacterial chitinase from Serratia marcescens.
Potato	Erwinia carotovora	Bacteriophage T4 lysozyme gene.
Tobacco	Botrytis cinerea	Stilbene (Phytoalexin-resveratol) synthase gene (STS) from peanut.
Tobacco	Tobacco Mosaic Virus	Coat protein gene of TMV (interfere with uncoating; virus spread).
Canola (Roundup Ready)	Herbicide Glyphosate	aroA (Agrobacterium sp.) encode 5-enolpyruvylshikimate-3-phosphate synthase (EPSPS) enzyme.
Cotton (BXN)	Bromoxynil	bxn gene (Klebsiella ozaene) converts herbicide into 3,5-dibromo4-hydroxy benzoic acid.

Potential Improvements in Nutritive Value of Plants

In an attempt to get better human health, mainly in immature countries, balancing of human nutrition has been carried out by supplying balanced sources of nutrition by creating hereditarily distorted foods that hold nutrients known to help fight disease or starvation. The major component of human diet is oil and fat, in which saturated fatty acid compounds are present, which are hazardous to health. GM foods have been developed with reduced levels of saturated fatty acid components in soybean and castor.

Improved Crop Productivity

A lesser quantity of than 20% of the earth is arable land but some crops have been hereditarily altered to make them more liberal of conditions like salinity, cold and drought. The detection of genes in plants in charge for sodium uptake has lead to growth of knock-out plants able to grow in high salt environments. Up- or down-regulation of record is usually the method used to alter drought-tolerance in plants. A global perspective suggests that some progress toward increased productivity has been made as insect-resistant, draught resistant and herbicide-tolerant transgenic varieties are reducing the risk of crop losses. One of the developments is the identification of a plant gene, At-DBF2, from Arabidopsis thaliana. Arabidopsis thaliana (tiny weed) shows tolerance to salt, drought and the heat and cold in plants. When this gene was inserted into tomato and tobacco cells, these cells withstood these conditions far better than ordinary cells similarly, Pizzuti and Daroda, also investigated the recombinant proteins from the root exudates of transformed tobacco for the major environmental concerns.

Improved Resistance to Pesticides/Herbicides

Intensive agriculture adversely affects the environment because farmers are using chemical inputs for optimising soil nutrient conditions and pesticides for controlling insects, pathogens and weeds. As pesticide resistant plants are broadminded of pesticides, allow farmers to selectively kill nearby weeds with no harming their crop. The most well-known example of this is the Roundup-Ready technology, urbanized by Monsanto. First introduced in 1998 as GM soybeans, Roundup-Ready plants are unaffected by the herbicide glyphosate, which can be applied in copious quantity to get rid of any other plants in the field. The decreased requirement of herbicides (10 - 40%) on herbicide tolerant varieties results in better control of weeds, improved yields, no-carryover of herbicide residues, better soil, water and air quality with savings in time and costs associated with conservative tillage to reduce weeds, or multiple applications of different types of herbicides to selectively eliminate exact species of weeds.

Improved Texture or Appearance of Food

Modified fruit can ripen longer on the plant and then be transported to the consumer with less risk of spoilage, and a still-reasonable shelf life. The first genetically modified food product was a tomato which was transformed to delay its ripening.

Better Flavour

Flavor can be altered by enhancing the activity of plant enzymes that transform aroma precursors into flavoring compounds. Transgenic peppers and melons with improved flavor are currently in field trials.

Insect/Pest Resistance

Most of the current commercial applications of modern biotechnology in agriculture are on reducing the dependence of farmers on agrochemicals. Insect/pest resistance crop provide more effective targeting of insects protected within plants for longer season protection. At the same time reduces the pesticide use and their exposure to farmers, labourers and non-target organisms e.g. Bt cotton: Engineered to contain and express the genes for Bt toxin, which they produce in its active form. When a susceptible insect ingests the transgenic crop cultivar expressing the Bt protein, it stops feeding and soon thereafter dies as a result of the Bt toxin binding to its gut wall. Bt corn: Engineered to control corn borer (a lepidopteran insect), which is otherwise controlled by spraying (a more difficult process).

Disease Resistance

Biotechnology has helped to increase crop productivity by introducing qualities as disease resistance. For example, researchers from the University of Hawaii and Cornell University developed two varieties of papaya resistant to papaya ringspot virus by transferring one of the virus' genes to papaya to create resistance in the plants. Seeds of the two varieties, named 'SunUp' and 'Rainbow', have been distributed under licensing agreements to papaya growers since 1998.

Transgenic Crops: For Molecular Farming

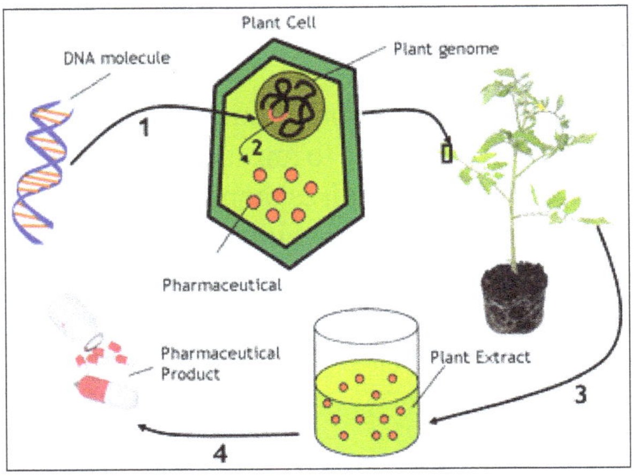

Plant Molecular Farming (PMF) consists of using transgenic plants as production platforms for the synthesis of compounds for pharmaceutical or industrial purposes. Insulin, produced by bacteria engineered with recombinant human DNA, was the first biopharmaceutical to be introduced to the market in 1982. In 1989 reported production of plantibody IgG. The first recombinant plant-derived pharmaceutical protein (PDP) was human serum albumin, initially produced in 1990 in transgenic tobacco and potato plants.

Vaccines

Oral vaccines have been in the works for much existence as a likely solution to the increase of disease in immature countries, where costs are excessive to extensive vaccination. Hereditarily engineered crops, frequently fruits or vegetables, planned to carry antigenic proteins from transferable pathogens that will activate an immune reply when injected. An example of this is a patientspecific vaccine for treating cancer. An anti-lymphoma vaccine has been made using tobacco plants carrying RNA from cloned malignant B-cells. The resultant protein is then used to vaccinate the patient and boost their immune system beside the cancer. Plant made vaccine: Transgenic plants are developed to express the antigens. Human or animal vaccines and expressed by transgenic plants:

Origin	Plant	Recombinant protein
Enterotoxigenic E. coli	Tobacco/potato	Heat-labile enterotoxin B
Vibrio cholera	Potato	Cholera CtoxA and Ctox B subunits
Hepatitis B virus	Potato/lettuce	Envelope surface protein
Rabies virus	Tomato	Rabies virus glycoprotein
Foot-and mouth disease	Arabidopsis/ alfalfa	Virus epitope VP1

Antibiotics

Plants are used to create antibiotics for both human and animal use. An expressing antibiotic protein in stock feed, fed straight to animals, is less expensive than traditional antibiotic production, but this practice raises many bioethics issues, because the result is widespread, possibly needless use of antibiotics which may encourage expansion of antibiotic-resistant bacterial strain. Quite a few rewards to using plants to create antibiotics for humans are condensed costs due to the larger quantity of product that can be produced from plants versus a fermentation unit, ease of purification, and condensed risk of contamination compared to that of using mammalian cells and culture media.

Production of plantibodies (antibody produced by genetically modified crops):

Plant	Antibody	Application
Tobacco	sIgA (hybrid)	S. mutans (dental caries)
Tobacco	IgG (guy's 13)	S. mutans (dental caries)
Tobacco	IgG Co 17-1A	Surface antigen(colon cancer)
Soybean	IgG (anti HSV-2)	Herpes simplex virus
Tobacco	scFv (38C13)	Lymphoma

Production of pharmaceuticals: Following table represents range of proteins of pharmaceutical interest that have been expressed in plants.

Recom binant protein	Origin (gene)	Plant (protein production)	Application
Protein C	Human	Tobacco	Anticoagulant
Hirudin	Hiruda medicinalis	Canola	Anticoagulant
Somatotrophin	Human	Tobacco	Growth hormone
Beta-Interferon	Human	Tobacco	Treatment for hepatitis B + C
Serum albumin	Human	Tobacco	Burns/fluid replacement
Glucocerebrosidase	Human	Tobacco	Gaucher's disease
α1 Antitrypsin	Human	Rice	Cystic fibrosis, haemorrhages
Aprotinin	Human	Maize	Transplant surgery
Trichosanthin-α	Trichosanthes kirilowii	Tobacco	HIV therapy, cancer

Production of Enzymes

Following table summarises recombinant enzyme production in plants:

Enzyme	Use
Avidin	Diagnostic kits
β- Glucuronidase	Diagnostic kits
Trypsin	Pharmaceuticals, wound care
Cellulase	Ethanol production from cellulose waste
Xylanase	Biomass processing
Phytase	Phytase breakdown, improved phosphate utilisation
α- Amylase	Food processing

(1-3) (1-4) β- Glucanase	Brewing
Lignin peroxidise	Paper manufacture

Production of carbohydrates, lipids and their derivatives:

Compound	Origin of genes	Origin of genes	Transgenic plant
Amylose –free starch	Potato	Food, industrial	Potato
Cyclodextrins	Klebsiella pneumonia	Food, pharmaceutical	Potato
Polyhyroxybutyrate	Klebsiella pneumonia	Biodegradable plastics	Soybean

2

Plant Tissue Culture

The culturing of plant seeds, organs, explants, tissues, cells or protoplasts on chemically defined synthetic nutrient media under sterile and controlled conditions is known as plant tissue culture. There are various methods of plant tissue culture such as seed culture, embryo culture, callus culture, organ culture, etc. This chapter discusses plant tissue culture and its different types in detail.

Most methods of plant transformation applied to GM crops require that a whole plant is regenerated from isolated plant cells or tissue which has been genetically transformed. This regeneration is conducted in vitro so that the environment and growth medium can be manipulated to ensure a high frequency of regeneration. In addition to a high frequency of regeneration, the re-generable cells must be accessible to gene transfer by whatever technique is chosen. The primary aim is therefore to produce, as easily and as quickly as possible, a large number of re-generable cells that are accessible to gene transfer. The subsequent regeneration step is often the most difficult step in plant transformation studies. However, it is important to remember that a high frequency of regeneration does not necessarily correlate with high transformation efficiency.

The first true cultures were obtained by Gautheret from cambial tissue of Acer pseudoplatanus. The term plant tissue culture (Micro propagation) is generally used for the aseptic culture of cells, tissues, organs and their components under defined chemical and physical conditions in vitro. The basic concept of the plant body can be dissected into smaller part termed as "explants" and any explants can be developed into a whole plant. It is a central innovative area of applied plant science, including agriculture and plant biotechnology. This technique is effective because almost all the plants cell is totipotent; in each cell possesses the genetic information and cellular machinery necessary to generate the whole organism. Since, this technique can be used to produce a higher number of plants that are genetically similar to a parent plant as well as to another.

Two concepts, plasticity and totipotency, are the central processes to understand the regeneration and plant cell culture. Plants, due to its longer life span and sessile nature, have developed a greater ability to overcome the extreme conditions. Most of the processes included in plant development and the growth, adapt to environmental conditions. When the plant cells and tissues are cultured in vitro, most of them are generally exhibit a very high degree of plasticity, which allows one type of organ or tissue to be initiated from another type. Like this way, the whole plant can be subsequently regenerated. This maintenance of genetic potential is called totipotency.

Tissue culture is the in vitro aseptic culture of cells, tissues, organs or whole plant under controlled nutritional and environmental conditions often to produce the clones of plants. The resultant clones are true-to type of the selected genotype. The controlled conditions provide the culture an environment conducive for their growth and multiplication. These conditions include proper supply of nutrients, pH medium, adequate temperature and proper gaseous and liquid environment.

Plant tissue culture technology is being widely used for large scale plant multiplication. Apart from their use as a tool of research, plant tissue culture techniques have in recent years, become of major industrial importance in the area of plant propagation, disease elimination, plant improvement and production of secondary metabolites.Small pieces of tissue (named explants) can be used to produce hundreds and thousands of plants in a continuous process. A single explant can be multiplied into several thousand plants in relatively short time period and space under controlled conditions, irrespective of the season and weather on a year round basis. Endangered, threatened and rare species have successfully been grown and conserved by micropropagation because of high coefficient of multiplication and small demands on number of initial plants and space.

In addition, plant tissue culture is considered to be the most efficient technology for crop improvement by the production of somaclonal and gametoclonal variants. The micropropagation technology has a vast potential to produce plants of superior quality, isolation of useful variants in well-adapted high yielding genotypes with better disease resistance and stress tolerance capacities. Certain type of callus cultures give rise to clones that have inheritable characteristics different from those of parent plants due to the possibility of occurrence of somaclonal variability, which leads to the development of commercially important improved varieties. Commercial production of plants through micropropagation techniques has several advantages over the traditional methods of propagation through seed, cutting, grafting and air-layering etc. It is rapid propagation processes that can lead to the production of plants virus free. Coryodalisyanhusuo, an important medicinal plant was propagated by somatic embryogenesis from tuber-derived callus to produce disease free tubers. Meristem tip culture of banana plants devoid from banana bunchy top virus (BBTV) and brome mosaic virus (BMV) were produced. Higher yields have been obtained by culturing pathogen free germplasmin vitro. Increase in yield up to 150% of virus-free potatoes was obtained in controlled conditions. The main objective of writing this chapter is to describe the tissue culture techniques, various developments, present and future trends and its application in various fields.

Basics of Plant Cell and Tissue Culture

In plant cell culture, plant tissues and organs are grown in vitro on artificial media, under aseptic and controlled environment. The technique depends mainly on the concept of totipotentiality of plant cells which refers to the ability of a single cell to express the full genome by cell division. Along with the totipotent potential of plant cell, the capacity of cells to alter their metabolism, growth and development is also equally important and crucial to regenerate the entire plant. Plant tissue culture medium contains all the nutrients required for the normal growth and development of plants. It is mainly composed of macronutrients, micronutrients, vitamins, other organic components, plant growth regulators, carbon source and some gelling agents in case of solid medium. Murashige and Skoog medium (MS medium) is most extensively used for the vegetative propagation of many plant species in vitro. The pH of the media is also important that affects both the growth of plants and activity of plant growth regulators. It is adjusted to the value between 5.4 - 5.8. Both the solid and liquid medium can be used for culturing. The composition of the medium, particularly the plant hormones and the nitrogen source has profound effects on the response of the initial explant.

Plant growth regulators (PGR's) play an essential role in determining the development pathway of plant cells and tissues in culture medium. The auxins, cytokinins and gibberellins are most commonly used plant growth regulators. The type and the concentration of hormones used depend mainly on the species of the plant, the tissue or organ cultured and the objective of the experiment. Auxins and cytokinins are most widely used plant growth regulators in plant tissue culture and their amount determined the type of culture established or regenerated. The high concentration of auxins generally favors root formation, whereas the high concentration of cytokinins promotes shoot regeneration. A balance of both auxin and cytokinin leads to the development of mass of undifferentiated cells known as callus.

Maximum root induction and proliferation was found in Stevia rebaudiana, when the medium is supplemented with 0.5 mg/l NAA. Cytokinins generally promote cell division and induce shoot formation and axillary shoot proliferation. High cytokinin to auxin ratio promotes shoot proliferation while high auxin to cytokinins ratio results in root formation. Shoot initiation and proliferation was found maximum, when the callus of black pepper was shifted to medium supplemented with BA at the concentration of 0.5 mg/l. Gibberellins are used for enhanced growth and to promote cell elongation. Maximum shoot length was observed in Phalaenopsis orchids when cultured in medium containing 0.5 mg/l GA3 (unpublished).

Tissue Culture in Agriculture

As an emerging technology, the plant tissue culture has a great impact on both agriculture and industry, through providing plants needed to meet the ever increasing world demand. It has made significant contributions to the advancement of agricultural sciences in recent times and today they constitute an indispensable tool in modern agriculture.

Biotechnology has been introduced into agricultural practice at a rate without precedent. Tissue culture allows the production and propagation of genetically homogeneous, disease-free plant material. Cell and tissue in vitro culture is a useful tool for the induction of somaclonal variation. Genetic variability induced by tissue culture could be used as a source of variability to obtain new stable genotypes. Interventions of biotechnological approaches for in vitro regeneration, mass micropropagation techniques and gene transfer studies in tree species have been encouraging. In vitro cultures of mature and/or immature zygotic embryos are applied to recover plants obtained from inter-generic crosses that do not produce fertile seeds. Genetic engineering can make possible a number of improved crop varieties with high yield potential and resistance against pests. Genetic transformation technology relies on the technical aspects of plant tissue culture and molecular biology for:

- Production of improved crop varieties.
- Production of disease-free plants (virus).
- Genetic transformation.
- Production of secondary metabolites.
- Production of varieties tolerant to salinity, drought and heat stresses.

Germplasm Conservation

In vitro cell and organ culture offers an alternative source for the conservation of endangered genotypes. Germplasm conservation worldwide is increasingly becoming an essential activity due to the high rate of disappearance of plant species and the increased need for safeguarding the floristic patrimony of the countries. Tissue culture protocols can be used for preservation of vegetative tissues when the targets for conservation are clones instead of seeds, to keep the genetic background of a crop and to avoid the loss of the conserved patrimony due to natural disasters, whether biotic or abiotic stress. The plant species which do not produce seeds (sterile plants) or which have 'recalcitrant' seeds that cannot be stored for long period of time can successfully be preserved via in vitro techniques for the maintenance of gene banks.

Cryopreservation plays a vital role in the long-term in vitro conservation of essential biological material and genetic resources. It involves the storage of in vitro cells or tissues in liquid nitrogen that results incryo-injury on the exposure of tissues tophysical and-chemical stresses. Successful cryopreservation is often ascertained by cell and tissue survival and the ability to re-grow or regenerate into complete plants or form new colonies. It is desirable to assess the genetic integrity of recovered germplasm to determine whether it is 'true-to-type' following cryopreservation. The fidelity of recovered plants can be assessed at phenotypic, histological, cytological, biochemical and molecular levels, although, there are advantages and limitations of the various approaches used to assess genetic stability. Cryobionomics is a new approach to study genetic stability in the cryopreserved plant materials. The embryonic tissues can be cryopreserved for future use or for germplasm conservation.

Embryo Culture

Embryo culture is a type of plant tissue culture that is used to grow embryos from seeds and ovules in a nutrient medium. In embryo culture, the plant develops directly from the embryo or indirectly through the formation of callus and then subsequent formation of shoots and roots. The technique has been developed to break seed dormancy, test the vitality of seeds, production of rare species and haploid plants. It is an effective technique that is employed to shorten the breeding cycle of plants by growing excised embryos and results in the reduction of long dormancy period of seeds. Intra-varietal hybrids of an economically important energy plant "Jatropha" have been produced successfully with the specific objective of mass multiplication. Somatic embryogenesis and plant regeneration has been carried out in embryo cultures of Jucara Palm for rapid cloning and improvement of selected individuals. In addition, conservation of endangered species can also be attained by practicing embryo culture technique. Recently a successful protocol has been developed for the in vitro propagation of Khayagrandifoliola by excising embryos from mature seeds. The plant has a high economic value for timber wood and for medicinal purposes as well. This technique has an important application in forestry by offering a mean of propagation of elite individuals where the selection and improvement of natural population is difficult.

Genetic Transformation

Genetic transformation is the most recent aspect of plant cell and tissue culture that provides the mean of transfer of genes with desirable trait into host plants and recovery of transgenic plants. The technique has a great potential of genetic improvement of various crop plants by integrating in plant biotechnology and breeding programmes. It has a promising role for the introduction of agronomically important traits such as increased yield, better quality and enhanced resistance to pests and diseases.

Genetic transformation in plants can be achieved by either vector-mediated (indirect gene transfer) or vectorless (direct gene transfer) method. Among vector dependant gene transfer methods, Agrobacterium-mediated genetic transformation is most widely used for the expression of foreign genes in plant cells. Successful introduction of agronomic traits in plants was achieved by using root explants for the genetic transformation. Virus-based vectors offers an alternative way of stable and rapid transient protein expression in plant cells thus providing an efficient mean of recombinant protein production on large scale.

Recently successful transgenic plants of Jatropha were obtained by direct DNA delivery to mature seed-derived shoot apices via particle bombardment method. This technology has an important impact on the reduction of toxic substances in seeds thus overcoming the obstacle of seed utilization in various industrial sector. Regeneration of disease or viral resistant plants is now achieved by employing genetic transformation technique. Researchers succeeded in developing transgenic plants of potato resistant to potato virus Y (PVY) which is a major threat to potato crop worldwide. In addition, marker free transgenic plants of Petunia hybrida were produced using multi-auto-transformation (MAT) vector system. The plants exhibited high level of resistance to Botrytis cinerea,causal agent of gray mold.

Protoplast Fusion

Somatic hybridization is an important tool of plant breeding and crop improvement by the production of interspecific and intergeneric hybrids. The technique involves the fusion of protoplasts of two different genomes followed by the selection of desired somatic hybrid cells and regeneration of hybrid plants. Protoplast fusion provides an efficient mean of gene transfer with desired trait from one species to another and has an increasing impact on crop improvement. Somatic hybrids were produced by fusion of protoplasts from rice and ditch reed using electrofusion treatment for salt tolerance.

In vitro fusion of protoplast opens a way of developing unique hybrid plants by overcoming the barriers of sexual incompatibility. The technique has been applicable in horticultural industry to create new hybrids with increased fruit yield and better resistance to diseases. Successful viable hybrid plants were obtained when protoplasts from citrus were fused with other related citrinae species. The potential of somatic hybridization in important crop plants is best illustrated by the production of intergeneric hybrid plants among the members of Brassicaceae. To resolvethe problem of loss of chromosomes and decreased regeneration capacity, successful protocol has been established for the production of somatic hybrid plants by using two types of wheat protoplast as recipient and protoplast of Haynaldiavillosa as a fusion donor. It is also employed as an important gene source for wheat improvement.

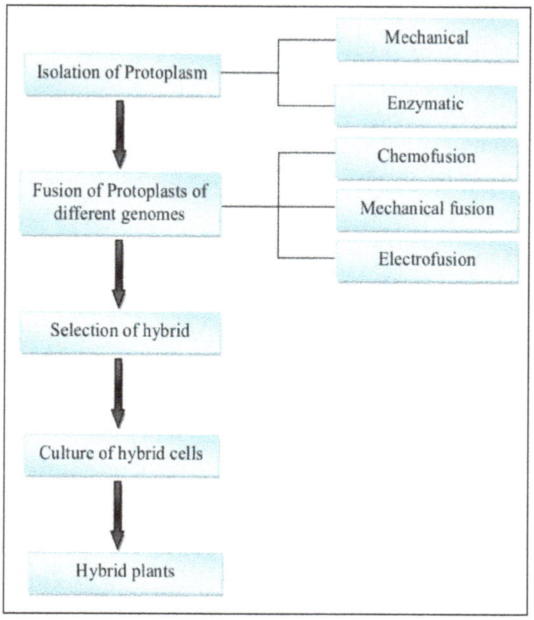

Figure: Schematic representation of production of hybrid plant via protoplast fusion.

Haploid Production

The tissue culture techniques enable to produce homozygous plants in relatively short time period through the protoplast, anther and microspore cultures instead of conventional breeding. Haploids are sterile plants having single set of chromosomes which are converted into homozygous diploids by spontaneous or induced chromosome doubling. The doubling of chromosomes restores the fertility of plants resulting in production of double haploids with potential to become pure breeding new cultivars. The term androgenesis refers to the production of haploid plants from young pollen cells without undergoing fertilization. Sudherson et al. reported haploid plant production of sturt's desert pea by using pollen grains as primary explants. The haploidy technology has now become an integral part of plant breeding programs by speeding up the production of inbred lines and overcoming the constraints of seed dormancy and embryo non-viability. The technique has a remarkable use in genetic transformation by the production of haploid plants with induced resistance to various biotic and abiotic stresses. Introduction of genes with desired trait at haploid state followed by chromosome doubling led to the production of double haploids inbred wheat and drought tolerant plantswere attained successfully.

Tissue Culture in Pharmaceuticals

Plant cell and tissue cultures hold great promise for controlled production of myriad of useful secondary metabolites. Plant cell cultures combine the merits of whole-plant systems with those of microbial and animal cell cultures for the production of valuable therapeutic secondary metabolites. In the search for alternatives to production of medicinal compounds from plants, biotechnological approaches, specifically plant tissue cultures, are found to have potential as a supplement to traditional agriculture in the industrial production of bioactive plant metabolites. Exploration of the biosynthetic capabilities of various cell cultures has been carried out by a group of plant scientists and microbiologists in several countries during the last decade.

Cell suspension culture: Cell suspension culture systems are used now days for large scale culturing of plant cells from which secondary metabolites could be extracted. A suspension culture is developed by transferring the relatively friable portion of the callus into liquid medium and is maintained under suitable conditions of aeration, agitation, light, temperature and other physical parameters. Cell cultures cannot only yield defined standard phytochemicals in large volumes but also eliminate the presence of interfering compounds that occur in the field-grown plants. The advantage of this method is that it can ultimately provide a continuous, reliable source of natural products. The major advantage of the cell cultures include synthesis of bioactive secondary metabolites, running in controlled environment, independently from climate and soil conditions.A number of different types of bioreactors have been used for mass cultivation of plant cells. The first commercial application of large scale cultivation of plant cells was carried out in stirred tank reactors of 200 liter and 750 liter capacities to produce shikonin by cell culture of Lithospermumerythrorhizon. Cell of Catharanthus-roseus, Dioscoreadeltoidea, Digitalis lanata, Panaxnotoginseng, Taxuswallichiana and Podophyllumhexandrum have been cultured in various bioreactors for the production of secondary plant products.

A number of medicinally important alkaloids, anticancer drugs, recombinant proteins and food additives are produced in various cultures of plant cell and tissues. Advances in the area of cell cultures for the production of medicinal compounds has made possible the production of a wide variety of pharmaceuticals like alkaloids, terpenoids, steroids, saponins, phenolics, flavanoids and amino acids. Some of these are now available commercially in the market for example shikonin and paclitaxel (Taxol). Until now 20 different recombinant proteins have been produced in plant cell culture, including antibodies, enzymes, edible vaccines, growth factors and cytokines. Advances in scale-up approaches and immobilization techniques contribute to a considerable increase in the number of applications of plant cell cultures for the production of compounds with a high added value. Some of the secondary plant products obtained from cell suspension culture of various plants are given in Table.

Table: List of some secondary plant product produced in suspension culture.

Secondary metabolite	Plant name
Vasine	*Adhatodavasica*
Artemisinin	*Artemisia annua*
Azadirachtin	*Azadirachtaindica*
Cathin	*Bruceajavanica*
Capsiacin	*Capsicum annum*
Sennosides	*Cassia senna*
Ajmalicine Secologanin Indole alkaloids Vincristine	*Catharanthusroseus*
Stilbenes	*Cayratiatrifoliata*
Berberin	*Cosciniumfenustratum*
Sterols	*Hyssopusofficinalis*
Shikonin	*Lithospermumerythrorhizon*
Ginseng saponin	*Panaxnotoginseng*
Podophyllotoxin	*Podophyllumhexandrum*
Taxane Paclitaxel	*Taxuschinensis*

Hairy Root Cultures

The hairy root system based on inoculation with Agrobacterium rhizogenes has become popular in the last two decades as a method of producing secondary metabolites synthesized in plant roots. Organized cultures, and especially root cultures, can make a significant contribution in the production of secondary metabolites. Most of the research efforts that use differentiated cultures instead of cell suspension cultures have focused on transformed (hairy) roots. Agrobacterium rhizogenes causes hairy root disease in plants. The neoplastic (cancerous) roots produced by A. rhizogenes infection are characterized by high growth rate, genetic stability and growth in hormone free media. High stability and productivity features allow the exploitation of hairy roots as valuable biotechnological tool for the production of plant secondary metabolites. These genetically transformed root cultures can produce levels of secondary metabolites comparable to that of intact plants. Hairy root technology has been strongly improved by increased knowledge of molecular mechanisms underlying their development. Optimizing the composition of nutrients for hairy root cultures is critical to gain a high production of secondary metabolites. Some of the secondary plant products obtained from hairy root culture of various plants are shown in Table.

Table: List of some secondary plant product produced in Hairy root culture.

Secondary metabolite	Plant name
Rosmarinic acid	*Agastacherugosa*
Deoursin	*Angelica gigas*
Resveratol	*Arachyshypogaea*
Tropane	*Brugmansia candida*
Asiaticoside	*Centellaasiatica*
Rutin	*Fagopyrumesculentum*
Glucoside	*Gentianamacrophylla*
Glycyrrhizin	*Glycyrrhizaglabra*
Shikonin	*Lithospermumerythrorhizon*
Glycoside	*Panax ginseng*
Plumbagin	*Plumbagozeylanica*
Anthraquinone	*Rubiaakane*
Silymarin	*Silybiummarianum*
Flavonolignan	*Silybiummariyanm*
Vincamine	*Vinca major*
Withanoloid A	*Withaniasomnifera*

Plant Growth Regulators and Tissue Culture

Generalizations about plant growth regulators and their use in plant cell culture media have been developed from initial observations made in the 1950s. There is, however, some considerable difficulty in predicting the effects of plant growth regulators: this is because of the great differences in culture response between species, cultivars and even plants of the same cultivar grown under different conditions. The following are the most commonly used plant growth regulators which are considered as critical media components in determining the developmental pathway of the plant cells.

- Auxins.
- Cytokinins.
- Gibberellins.
- Abscisic acid.
- Ethylene.

Auxins and cytokinins are the most widely used plant growth regulators in plant tissue culture and are usually used together, the ratio of the auxin to the cytokinin determining the type of culture established or regenerated. A high auxin to cytokinin ratio generally favors root formation, whereas a high cytokinin to auxin ratio favors shoot formation. An intermediate ratio favors callus production.

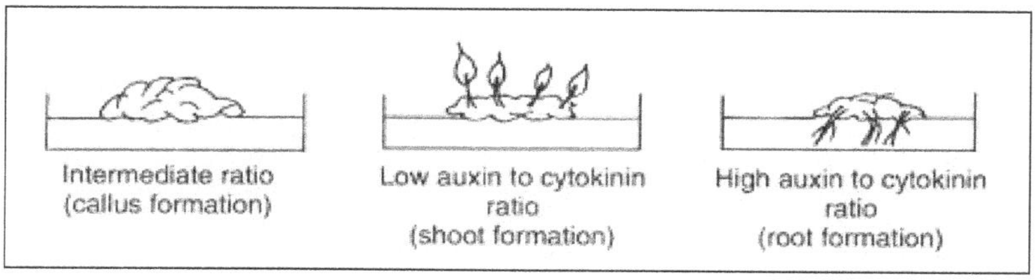

The effect of different ratios of auxin to cytokinin on the growth and morphogenesis of callus. High auxin to cytokinin ratios promote root development, low ratios promote shoot development. Intermediate ratios promote continued growth of the callus without differentiation.

Auxins are a group of phytohormones produced in the shoot and root apices and they migrate from the apex to the zone of elongation. Auxins promote the growth along the longitudinal axis of the plant and hence the name (auxeing : to grow). The term, auxin was introduced by Kogl and Haagen- Smit. Went isolated auxin from the Avena coleoptile tips by a method called Avena coleoptile or curvature test and concluded that no growth can occur without auxin. Auxins are widely distributed throughout the plant however, abundant in the growing tips such as coleoptile tip, buds, root tips and leaves. Indole Acetic Acid (IAA) is the only naturally occurring auxin in plants. The synthetic auxins include:

- IBA : Indole Butyric Acid.
- NAA : Naphthalene Acetic acid.
- MENA: Methyl ester of Naphthalene acetic acid.
- MCPA: 2 Methyl 4 chloro phenoxy acetic acid.
- TIBA : 2, 3, 5 Tri iodo benzoic acid.
- 2, 4-D : 2, 4 dichloro phenoxy acetic acid.

- 2, 4, 5-T: 2, 4, 5 – Trichloro phenoxy acetic acid.

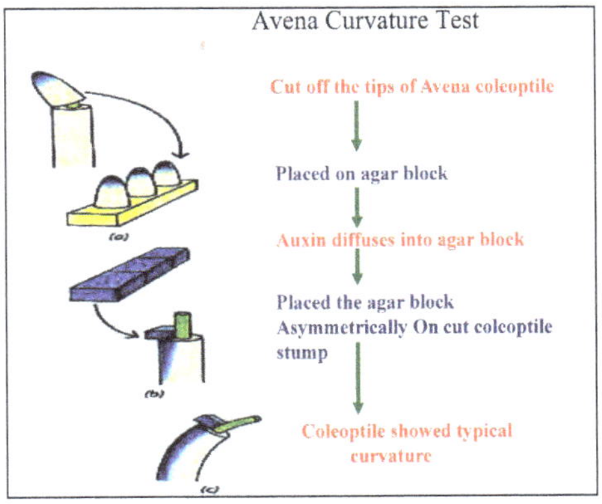

Natural auxins may occur in the form of either free auxins- which freely move or diffuse out of the plant tissues readily or bound auxins- which are released from plant tissues only after hydrolysis, autolysis or enzymolysis.

Physiological Effects of Auxin

Cell Division and Elongation

The primary physiological effects of auxin are cell division and cell elongation in the shoots. It is important in the secondary growth of stem and differentiation of xylem and phloem tissues.

Apical Dominance

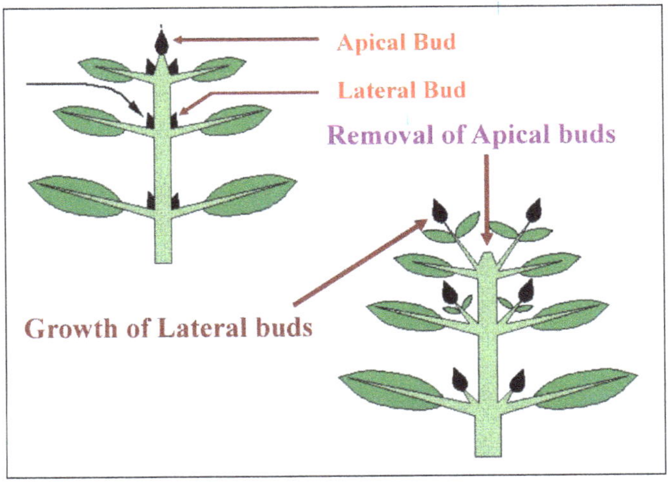

In many plants, if the terminal bud is intact and growing, the growth of lateral buds just below it remains suppressed. Removal of the apical bud results in the rapid growth of lateral buds. This phenomenon in which the apical bud dominates over the lateral buds and does not allow the lateral buds to grow is known as apical dominance.

Skoog and Thimmann pointed out that the apical dominance might be under the control of auxin produced at the terminal bud and which is transported downward through the stem to the lateral buds and hinders the growth. They removed the apical bud and replaced it with agar block. This resulted in rapid growth of lateral buds. But when they replaced the apical bud with agar block containing auxin, the lateral buds remained suppressed and did not grow.

Root Initiation

In contrast to stem, the higher concentration of auxin inhibits the elongation of roots but the number of lateral roots is considerably increased i.e., higher concentration of auxin induces more lateral branch roots. Application of IAA in lanolin paste (lanolin is a soft fat prepared from wool and is good solvent for auxin) to the cut end of a young stem results in an early and extensive rooting. This fact is of great practical importance and has been widely utilized to promote root formation in economically useful plants which are propagated by cuttings.

Prevention of Abscission

Natural auxins prevent the formation of abscission layer which may otherwise result in the fall of leaves, flowers and fruits.

Parthenocarpy

Auxin can induce the formation of parthenocarpic fruits (fruit formation without pollination and fertilization). In parthenocarpic fruits, the concentration of auxin in the ovaries is higher than in the ovaries of plants which produce fruits only after fertilization. In the later cases, the concentration of the auxin in ovaries increases after pollination and fertilization.

Respiration

Auxin stimulates respiration and there is a correlation between auxin induced growth and respiration. Auxin may increase the rate of respiration indirectly through increased supply of ADP by rapidly utilizing ATP in the expanding cells.

Callus Formation

Besides cell elongation, auxin may also be active in cell division. In many tissue cultures, where the callus growth is quite normal, the continued growth of such callus takes place only after the addition of auxin.

Eradication of Weeds

Some synthetic auxins especially 2, 4- D and 2, 4, 5-T are useful in eradication of weeds at higher concentrations.

Flowering and Sex Expression

Auxins generally inhibit flowering but in pine apple and lettuce it promotes uniform flowering.

Distribution of Auxin in Plants

In plants, auxin (IAA) is synthesized in growing tips or meristematic regions from where; it is transported to other plant parts. Hence, the highest concentration of IAA is found in growing shoot tips, young leaves and developing auxiliary shoots. In monocot seedling, the highest concentration of auxin is found in coleoptile tip which decreases progressively towards its base.

In dicot seedlings, the highest concentration is found in growing regions of shoot, young leaves and developing auxiliary shoots. Within the plants, auxin may present in two forms. i.e., free auxins and bound auxins. Free auxins are those which are easily extracted by various organic solvents such as diethyl ether. Bound auxins on the other hand, need more drastic methods such as hydrolysis, autolysis, enzymolysis etc. for extraction of auxin. Bound auxins occur in plants as complexes with carbohydrates such as glucose, arabionse or sugar alcohols or proteins or amino acids such as aspartate, glutamate or with inositol.

Biosynthesis of Auxin (IAA) in Plants

Thimann found that an amino acid, tryptophan is converted into Indole 3 acetic acid. Tryptophan is the primary precursor of IAA in plants. IAA can be formed from tryptophan by two different pathways:

- By deamination of tryptophan to form indole-3-pyruvic acid followed by decarboxylation to from indole-3-acetaldehyde. The enzymes involved are tryptophan deamintion and indole pyruvate decarboxylase respectively.
- By decarboxylation of tryptophan to form tryptamine followed by deamination to form indole-3-acetaldehyde and the enzymes involved are tryptophan decarboxylase and tryptamine oxidase respectively. Indole 3-acetaldehyde can readily be oxidized to indole 3-acetic acid (IAA) in the presence of indole 3-acetaldehyde dehydrogenase.

Transport of Auxin in Plant

The transport of auxin is predominantly polar. In stems, polar transport of auxin is basipetal i.e., it takes place from apex towards base. Polar transport of auxin is inhibited by 2, 3, 5 Triiodobenzoic acid (TIBA) and Naphthyl thalamic acid (NPA). The substances are called as antiauxins.

Destruction / Inactivation of Auxin in Plants

Auxin is destructed by the enzyme IAA oxidase in the presence of O_2 by oxidation.

$$IAA + H_2O_2 + O_2 \longrightarrow \text{3-methylene oxindole} + H_2O + CO_2$$

Rapid inactivation may also occur by irradiation with x-rays and gamma rays. UV light also reduces auxin levels in plants. Inactivation or decomposition of IAA by light has been called as photo oxidation.

Mechanism of Action

IAA increases the plasticity of cell walls so that the cells stretch easily in response to turgor pressure. It has been suggested that IAA acts upon DNA to influence the production of mRNA. The mRNA codes for specific enzymes responsible for expansion of cell walls. Recent evidences indicate that IAA increases oxidative phosphorylation in respiration and enhanced oxygen uptake. The growth stimulation might be due to increased energy supply and it is also demonstrated that auxin induces production of ethylene in plants.

Gibberellins

A Japanese scientist Kurosawa found that the rice seedlings infected by the fungus Gibberella fujikuroi grow taller and turned very thin and pale. An active substance was isolated from the infected seedlings and named as Gibberellin.

Biosynthesis of Gibberellins in Plants

The primary precursor for the formation of gibberellins is acetate:

Acetate + COA → Acetyl COA → Mevalonic acid → MA pyrophosphate → Isopentanyl pyrophosphate → Geranyl pyrophosphate → GGPP → Kaurene → Gibberellins.

Physiological Effects of Gibberellins

Seed Germination

Certain light sensitive seeds eg. Lettuce and tobacco show poor germination in dark. Germination starts vigorously if these seeds are exposed to light or red light. This requirement of light is overcome if the seeds are treated with gibberellic acid in dark.

Dormancy of Buds

In temperature regions the buds formed in autumn remain dormant until next spring due to severe cold. This dormancy of buds can be broken by gibberellin treatments. In potato also, there is a dormant period after harvest, but the application of gibberellin sprouts the refer vigorously.

Root Growth

Gibberellins have little or no effect on root growth. At higher concentration, some inhibition of root growth may occur. The initiation of roots is markedly inhibited by gibberellins in isolated cuttings.

Elongation of Internodes

The most pronounced effect of gibberellins on the plant growth is the elongation of the internodes. Therefore in many plants such as dwarf pea, dwarf maize etc gibberellins overcome the genetic dwarfism.

Bolting and Flowering

In many herbaceous plants, the early period of growth shows rosette habit with short stem and small leaves. Under short days, the rosette habit is retained while under long days bolting occurs i.e. the stem elongates rapidly and is converted into polar axis bearing flower primordia. This bolting can also be induced in such plants by the application of gibberellins even under non-inductive short days.

In Hyoscyamus niger (a long day plant) gibberellin treatment causes bolting and flowering under non-inductive short days. While in long day plants the gibberellin treatment usually results in early flowering. In short day plants, its effects are quite variable. It may either have no effect or inhibit or may activate flowering.

Parthenocarpy

Germination of the pollen grains is stimulated by gibberellins; likewise, the growth of the fruit and the formation of parthenocarpic fruits can be induced by gibberellin treatment. In many cases, eg. pome and stone fruits where auxins have failed to induce parthenocarpy, the gibberellins have proven to be successful. Seedless and fleshly tomatoes and large sized seedless grapes are produced by gibberellin treatments on commercial scale.

Synthesis of the Enzyme α – Amylase

One important function of gibberellins is to cause the synthesis of the enzyme α - amylase in the aleurone layer of the endosperm of cereal grains during germination. This enzyme brings about hydrolysis of starch to form simple sugars which are then translocated to growing embryo to provide energy source.

Distribution of Gibberellins in Plant

Gibberellins are found in all parts of higher plants including shoots, roots, leaves, flower, petals, anthers and seeds. In general, reproductive parts contain much higher concentrations of gibberellins than the negative parts. Immature seeds are especially rich in gibberellins (10-100 mg per g fresh weight). In plants, gibberellins occur in two forms free gibberellins and bound gibberellins. Bound gibberellins usually occur as gibberellin – glycosides.

Cytokinins (Kinetin)

Kinetin was discovered by Skoog and Miller from the tobacco pith callus and the chemical substance was identified as 6-furfuryl aminopurine. Because of its specific effect on cytokinesis (cell division), it was called as cytokinins or kinetin. The term, cytokinin was proposed by Letham. Fairley and Kingour used the term, phytokinins for cytokinins because of their plant origin. Chemically cytokinins are kinins and they are purine derivatives.

Cytokinins, besides their main effect on cell division, also regulate growth and hence they are considered as natural plant growth hormones. Some of the very important and commonly known naturally occurring cytokinins are Coconut milk factor and Zeatin. It was also identified that cytokinin as a constituent of t-RNA.

Naturally Occurring Cytokinins

Cytokinins can be extracted from coconut milk (liquid endosperm of coconut), tomato juice, flowers and fruits of Pyrus malus; fruits of Pyrus communis (Pear), Prunus cerasiferae (plum) and Lycopersicum esculentum (bhendi); Cambial tissues of Pinus radiata, Eucalyptus regnans and Nicotiana tabacum; immature fruits of Zea mays, Juglans sp. and Musa sp; female gametophytes of Ginkgo biloba; fruitlets, embryo and endosperms of Prunus persica; seedling of Pisum sativum; root exudates of Helianthus annuus and tumour tissues of tobacco. According to Skoog and Armstrong, at least seven well established types of cytokinins have been reported from the plants.

Biosynthesis

It is assumed that cytokinins are synthesised as in the case of purines in plants (nucleic acid synthesis). Root tip is an important site of its synthesis. However, developing seeds and cambial tissues are also the site of cytokinin biosynthesis. Kende reported that cytokinins move upwards perhaps in the xylem stream. However, basipetal movement in petiole and isolated stems are also observed. Seth et al found that auxin enhances kinetin movement (translocation) in bean stems.

Physiological Effects of Cytokinins

Cell Division

The most important biological effect of kinetin on plants is to induce cell division especially in tobacco pith callus, carrot root tissue, soybean cotyledon, pea callus etc.

Cell Enlargement

Like auxins and gibberellins, the kinetin may also induce cell enlargement. Significant cell enlargement has been observed in the leaves of Phaseolus vulgaris, pumpkin cotyledons, tobacco pith culture, cortical cells of tobacco roots etc.

Concentration of Apical Dominance

External application of cytokinin promotes the growth of lateral buds and hence counteracts the effect of apical dominance.

Dormancy of Seeds

Like gibberellins, the dormancy of certain light sensitive seeds such as lettuce and tobacco can also be broken by kinetin treatment.

Delay of Senescence (Richmand - Lang Effect)

The senescence of leaves usually accompanies with loss of chlorophyll and rapid breakdown of proteins. Senescence can be postponed to several days by kinetin treatment by improving RNA synthesis followed by protein synthesis. Richmand and Lang while working on detached leaves of Xanthium found that kinetin was able to postpone the senescence for a number of days.

Flower Induction

Cytokinins can be employed successfully to induce flowering in short day plants.

Morphogenesis

It has been shown that high auxin and low kinetin produced only roots whereas high kinetin and low auxin could promote formation of shoot buds.

Accumulation and Translocation of Solutes

Plants accumulate solutes very actively with the help of Cytokinin and also help in solute translocation in phloem.

Protein Synthesis

Osborne demonstrated the increased rate of protein synthesis due to translocation bys kinetin treatment.

Other Effects

Cytokinins provide resistance to high temperature, cold and diseases in some plants. They also help in flowering by substituting the photoperiodic requirements. In some cases, they stimulate synthesis of several enzymes involved in photosynthesis.

Commercial Applications

Cytokinins have been used for increasing shelf life of fruits, quickening of root induction and producing efficient root system, increasing yield and oil contents of oil seeds like ground nut.

Ethylene

Ethylene is the only natural plant growth hormone exists in gaseous form.

Important Physiological Elects

- The main role of ethylene is it hastens the ripening of fleshy fruits eg. Banana, apples, pears, tomatoes, citrus etc.
- It stimulates senescence and abscission of leaves.
- It is effective in inducing flowering in pine apple.
- It causes inhibition of root growth.
- It stimulates the formation of adventitious roots.
- It stimulates fading of flowers.
- It stimulates epinasty of leaves.

Abscisic Acid

Addicott isolated a substance strongly antagonistic to growth from young cotton fruits and named Abscissin II. Later on this name was changed to Abscisic acid. This substance also induces dormancy of buds therefore it also named as Dormin. Abscisic acid is a naturally occurring growth inhibitor.

Physiological Effects

The two main physiological effects are:

- Geotropism in roots.
- Stomatal closing.
- Besides other effects.

Geotropism In Roots

Geotropic curvature of root is mainly due to translocation of ABA in basipetal direction towards the root tip.

Stomatal Closing

ABA is synthesized and stored in mesophyll chloroplast. In respond to water stress, the permeability of chloroplast membrane is lost which resulted is diffusion of ABA out of chloroplast into the cytoplasm of the mesophyll cells. From mesophyll cells it diffuses into guard cells where it causes closing of stomata.

Other Effects

- Including bud dormancy and seed dormancy.
- Includes tuberisation.
- Induces senescence of leaves fruit ripening, abscission of leaves, flowers and fruits.
- Increasing the resistance of temperate zone plants to frost injury.

Growth Retardants

There is no. of synthesis compounds which prevent the gibberellins from exhibiting their usual responses in plants such as cell enlargement or stem elongation. So they are called as anti gibberellins or growth retardants. They are:

- Cycocel (2- chloroethyl trimethyl ammonium chloride (CCC).
- Phosphon D – (2, 4 – dichlorobenzyl – tributyl phosphonium chloride).
- AMO – 1618.
- Morphactins.
- Maleic hydrazide.

Plant Cell Culture Media

Culture media used for the cultivation of plant cells in vitro are composed of three basic components:

- Essential elements, or mineral ions, supplied as a complex mixture of salts.
- An organic supplement supplying vitamins and/or amino acids.
- 3A source of fixed carbon; usually supplied as the sugar sucrose.

For practical purposes, the essential elements are further divided into the following categories:

- Macroelements (or macronutrients).
- Microelements (or micronutrients).
- An iron source.

Optimal growth and morphogenesis of tissues may vary for different plants according to their nutritional requirements. Moreover, tissues from different parts of plants may also have different requirements for satisfactory growth. Tissue culture media were first developed from nutrient solutions used for culturing whole plants e.g. root culture medium of White and callus culture medium of Gautheret. White's medium was based on Uspenski and Uspenska's medium for algae, Gautheret's medium was based on Knop's salt solution. Basic media that are frequently used include Murashige and Skoog (MS) medium, Linsmaier and Skoog (LS) medium, Gamborg (B5) medium and Nitsch and Nitsch (NN) medium.

Media Composition

Plant tissue culture media should generally contain some or all of the following components: macronutrients, micronutrients, vitamins, amino acids or nitrogen supplements, source(s) of carbon, undefined organic supplements, growth regulators and solidifying agents. According to the International Association for Plant Physiology, the elements in concentrations greater than 0.5 mM.l-1 are defined as macroelements and those required in concentrations less than 0.5 mM.l-1 as microelements. It should be considered that the optimum concentration of each nutrient for achieving maximum growth rates varies among species.

Macronutrients

The essential elements in plant cell or tissue culture media include, besides C, H and O, macroelements: nitrogen (N), phosphorus (P), potassium (K), calcium (Ca), magnesium (Mg) and sulphur (S) for satisfactory growth and morphogenesis. Culture media should contain at least 25-60 mM of inorganic nitrogen for satisfactory plant cell growth. Potassium is required for cell growth of most plant species. Most media contain K in the form of nitrate chloride salts at concentrations ranging between 20 and 30 mM. The optimum concentrations of P, Mg, S and Ca range from 1-3 mM if other requirements for cell growth are provided.

Micronutrients

The essential micronutrients (minor elements) for plant cell and tissue growth include iron (Fe), manganese (Mn), zinc (Zn), boron (B), copper (Cu) and molybdenum (Mo). Iron is usually the most critical of all the micronutrients. The element is used as either citrate or tartarate salts in culture media, however, there exist some problems with these compounds for their difficulty to dissolve and precipitate after media preparation. There has been trials to solve this problem by using ethylene diaminetetraacetic acid (EDTA)-iron chelate (FeEDTA). A procedure for preparing an iron chelate solution that does not precipitate have been also developed. Cobalt (Co) and iodine (I) may be added to certain media, but their requirements for cell growth has not been precisely established. Sodium (Na) and chlorine (Cl) are also used in some media, in spite of reports that they are not essential for growth. Copper and cobalt are added to culture media at concentrations of 0.1μM, iron and molybdenum at 1μM, iodine at 5μM, zinc at 5-30 μM, manganese at 20-90 μM and boron at 25-100 μM.

Carbon and Energy Sources

In plant cell culture media, besides the sucrose, frequently used as carbon source at a concentration of 2-5%, other carbohydrates are also used. These include lactose, galactose, maltose and starch and they were reported to be less effective than either sucrose or glucose, the latter was similarly more effective than fructose considering that glucose is utilized by the cells in the beginning, followed by fructose. It was frequently demonstrated that autoclaved sucrose was better for growth than filter sterilized sucrose. Autoclaving seems to hydrolyze sucrose into more efficiently utilizable sugars such as fructose. Sucrose was reported to act as morphogenetic trigger in the formation of auxiliary buds and branching of adventitious roots.

It was found that supplements of sugar cane molasses, banana extract and coconut water to basal media can be a good alternative for reducing medium costs. These substrates in addition to sugars, they are sources of vitamins and inorganic ions required growth.

Vitamins and Myo-Inositol

Some plants are able to synthesize the essential requirements of vitamins for their growth. Some vitamins are required for normal growth and development of plants, they are required by plants as catalysts in various metabolic processes. They may act as limiting factors for cell growth and differentiation when plant cells and tissues are grown Invitro. The vitamins most used in the cell and tissue culture media include: thiamin (B1), nicotinic acid and pyridoxine (B6). Thiamin is necessarily required by all cells for growth. Thiamin is used at concentrations ranging from 0.1 to 10 mg.l-1. Nicotinic acid and pyridoxine, however not essential for cell growth of many species, they are often added to culture media. Nicotinic acid is used at a concentration range 0.1-5 mg.l-1 and pyridoxine is used at 0.1-10 mg.l-1. Other vitamins such as biotin, folic acid, ascorbic acid, pantothenic acid, tocopherol (vitamin E), riboflavin, p-amino-benzoic acid are used in some cell culture media however, they are not growth limiting factors. It was recommended that vitamins should be added to culture media only when the concentration of thiamin is below the desired level or when the cells are required to be grown at low population densities. Although it is not a vitamin but a carbohydrate, myo-inositol is added in small quantities to stimulate cell growth of most plant species. Myo-inositol is believed to play a role in cell division because of its breakdown to ascorbic acid and pectin and incorporation into phosphoinositides and phosphatidyl-inositol. It is generally used in plant cell and tissue culture media at concentrations of 50-5000 mg.l-1.

Amino Acids

The required amino acids for optimal growth are usually synthesized by most plants, however, the addition of certain amino acids or amino acid mixtures is particularly important for establishing cultures of cells and protoplasts. Amino acids provide plant cells with a source of nitrogen that is easily assimilated by tissues and cells faster than inorganic nitrogen sources. Amino acid mixtures such as casein hydrolysate, L-glutamine, L-asparagine and adenine are frequently used as sources of organic nitrogen in culture media. Casein hydrolysate is generally used at concentrations between 0.25-1 g.l-1. Amino acids used for enhancement of cell growth in culture media included; glycine at 2 mg.l-1, glutamine up to 8 mM, asparagine at 100mg.l-1, L-arginine and cysteine at 10 mg.l-1 and L-tyrosine at 100mg.l-1.

Undefined Organic Supplements

Some media were supplemented with natural substances or extracts such as protein hydrolysates, coconut milk, yeast extract, malt extract, ground banana, orange juice and tomato juice, to test their effect on growth enhancement. A wide variety of organic extracts are now commonly added to culture media. The addition of activated charcoal is sometimes added to culture media where it may have either a beneficial or deleterious effect. Growth and differentiations were stimulated in orchids, onions and carrots, tomatoes. On the other hand, an inhibition of cell growth was noticed on addition of activated charcoal to culture medium of soybean. Explanation of the mode of action of activated charcoal was based on adsorption of inhibitory compounds from the medium, adsorption of growth regulators from the culture medium or darkening of the medium. The presence of 1% activated charcoal in the medium was demonstrated to largely increase hydrolysis of sucrose during autoclaving which cause acidification of the culture medium.

Solidifying Agents

Figure: Agar-solidified medium supporting plant growth.

Hardness of the culture medium greatly influences the growth of cultured tissues. There are a number of gelling agents such as agar, agarose and gellan gum. Agar, a polysaccharide obtained from seaweeds, is of universal use as a gelling agent for preparing semi-solid and solid plant tissue culture media. Agar has several advantages over other gelling agents; mixed with water, it easily melts in a temperature range 60-100oC and solidifies at approximately 45oC and it forms a gel stable at all feasible incubation temperatures. Agar gels do not react with media constituents and are not digested by plant enzymes. It is commonly used in media at concentrations ranging between 0.8-1.0%. Pure agar preparations are of great importance especially in experiments dealing with tissue metabolism. Agar contains Ca, Mg and trace elements on comparing different agar brands; Bacto, Noble and purified agar, in concern with contaminants. For example reported Bacto agar to contain 0.13, 0.01, 0.19, 0.43, 2.54, 0.17% of Ca, Ba, Si, Cl, SO4-, N, respectively. Impurities also included 11.0, 285.0 and 5.0 mg.l1- for iron, magnesium and copper as contaminants, respectively. Amounts of some contaminants were higher in purified agar than in Bacto agar of which Mg that accounted for 695.0 mg.l1- and Cu for 20.0 mg.l1-.

Reduction of culture media costs is continually targeted in large-scale cultures and search for cheap alternatives provided that white flower, potato starch, rice powder were as good gelling agents as agar. It was also experienced that combination of laundry starch, potato starch and semolina in a ratio of 2:1:1 reduced costs of gelling agents by more than 70%.

Growth Regulators

Plant growth regulators are important in plant tissue culture since they play vital roles in stem elongation, tropism, and apical dominance. They are generally classified into the following groups; auxins, cytokinins, gibberellins and abscisic acid. Moreover, proportion of auxins to cytokinins determines the type and extent of organogenesis in plant cell cultures.

Auxins

The common auxins used in plant tissue culture media include: indole-3- acetic acid (IAA), indole-3- butricacide (IBA), 2,4-dichlorophenoxy-acetic acid (2,4-D) and naphthalene- acetic acid (NAA). IAA is the only natural auxin occurring in plant tissues There are other synthetic auxins used in culture media such as 4-chlorophenoxy acetic acid or p-chloro-phenoxy acetic acid (4-CPA, pCPA), 2,4,5-trichloro-phenoxy acetic acid (2,4,5 T), 3,6- dichloro-2-methoxy- benzoic acid (dicamba) and 4- amino-3,5,6-trichloro-picolinic acid (picloram).

Auxins differ in their physiological activity and in the extent to which they translocate through tissue and are metabolized. Based on stem curvature assays, eight to twelve times higher activity was reported on using 2,4-D than IAA, four times higher activity of 2,4,5 T in comparison with IAA and NAA has as doubled activity as IAA. In tissue cultures, auxins are usually used to stimulate callus production and cell growth, to initiate shoots and rooting, to induce somatic embryogenesis, to stimulate growth from shoot apices and shoot stem culture. The auxin NAA and 2,4-D are considered to be stable and can be stored at 40C for several months. The solutions of NAA and 2,4-D can also be stored for several months in a refrigerator or at -20°C if storage has to last for longer periods. It is best to prepare fresh IAA solutions each time during medium preparation, however IAA solutions can be stored in an amber bottle at 4°C for no longer than a week. Generally IAA and 2,4-D are dissolved in a small volume of 95% ethyl alcohol. NAA, 2,4-D and IAA can be dissolved in a small amount of 1N NaOH. Chemical structures of some of the frequently used auxins are given in Figure.

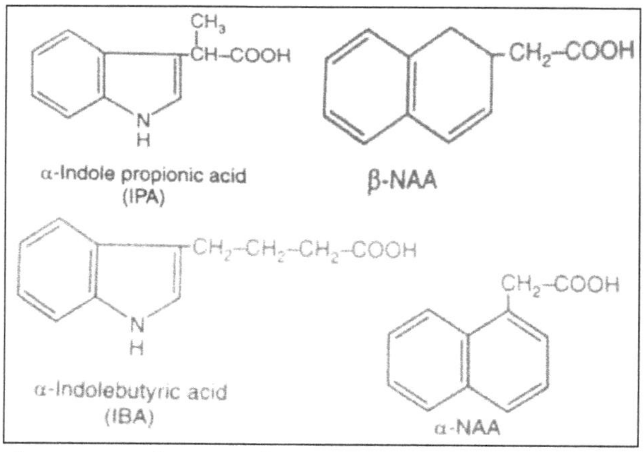

Figure: Chemical structure of commonly used auxins. IAA indole acetic acid, IBA Indole butyric acid, 2,4-D dichlorophenoxyacetic acid and NAA naphthalene acetic acid.

There are also some auxinlike compounds that vary in their activity and are rarely used in culture media.

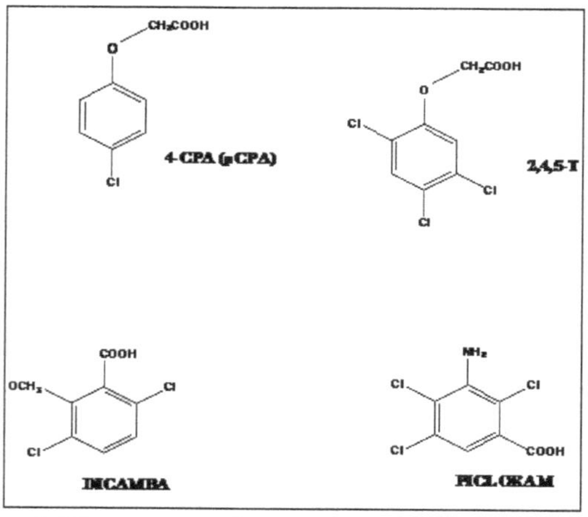

Figure: Some auxinlike compounds. pCPA, p-chloro-phenoxy acetic acid and 2,4,5T, 2,4,5-trichloro-phenoxy acetic acid.

Cytokinins

Cytokinins commonly used in culture media include BAP (6-benzyloaminopurine), 2iP

(6-dimethylaminopurine), kinetin (N-2-furanylmethyl-1H-purine-6-amine), Zeatin (6-4-hydroxy-3-methyl-trans-2-butenylaminopurine) and TDZ (thiazuron-N-phenyl-N-1,2,3 thiadiazol-5ylurea). Zeatin and 2iP are naturally occurring cytokinins and zeatin is more effective. In culture media, cytokinins proved to stimulate cell division, induce shoot formation and axillary shoot proliferation and to retard root formation. The cytokinins are relatively stable compounds in culture media and can be stored desiccated at -20 °C. Cytokinins are frequently reported to be difficult to dissolve and sometimes addition of few drops of 1NHCl or 1N NaOH facilitate their dissolution. Cytokinins can be dissolved in small amounts of dimethylsulfoxide (DMSO) without injury to the plant tissue. DMSO has an additional advantage because it acts as a sterilizing agent; thus stock solutions containing DMSO can be added directly to the sterile culture medium. Chemical structure of the frequently used in plant tissue culture media is given in figure.

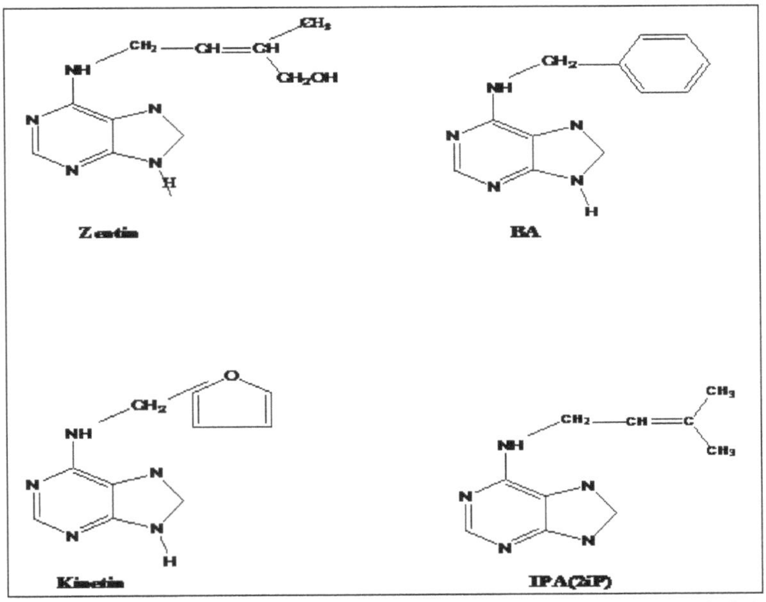

Figure: Chemical structure of some cytokinins, BA, benzyladenine, IPA dimethylallylamino purine.

Gibberellins

Gibberellins comprise more than twenty compounds, of which GA3 is the most frequently used gibberellin. These compounds enhance growth of callus and help elongation of dwarf plantlets.

Other growth regulators are sometimes added to plant tissue culture media as abscisic acid, a compound that is usually supplemented to inhibit or stimulate callus growth, depending upon the species. It enhances shoot proliferation and inhibits later stages of embryo development. Although growth regulators are the most expensive medium ingredients, they have little effect on the medium cost because they are required in very small concentrations.

A comparison of the chemical composition of the frequently used plant tissue culture media appears in Table which was given in the appendix of the proceedings of the Technical meeting of the International Atomic Energy Agency (IAEA).

Table: Composition of media most frequently used.

Medium Components ($mg.l^{1-}$)	MS	G_5	W	LM	VW	Km	M	NN
Macronutrients								
$Ca_3(PO4)_2$					200.0			
NH_4NO_3	1650.0			400.0				720.0
KNO_3	1900.0	2500.0	80.0		525.0	180.0	180.0	950.0
$CaCl_2.2H_2O$	440.0	150.0		96.0				166.0
$MgSO_4.7H_2O$	370.0	250.0	720.0	370.0	250.0	250.0	250.0	185.0
KH_2PO_4	170.0			170.0	250.0	150.0	150.0	68.0
$(NH4)_2SO4$		134.0			500.0	100.0	100.0	
$NaH_2PO_4.H_2O$		150.0	16.5					
$CaNO_3.4H_2O$			300.0	556.0		200.0	200.0	
Na_2SO_4			200.0					
KCl			65.0					
K_2SO_4				990.0				
Micronutrients								
KI	0.83	0.75	0.75			80.0	0.03	
H_3BO_3	6.20	3.0	1.5	6.2		6.2	0.6	10.0
$MnSO_4.4H_2O$	22.30		7.0		0.75	0.075		25.0
$MnSO_4.H_2O$		10.0		29.43				
$ZnSO_4.7H_2O$	8.6	2.0	2.6	8.6			0.05	10.0
$Na_2MoO_4.2H_2O$	0.25	0.25		0.25		0.25	0.05	0.25
$CuSO_4.5H_2O$	0.025	0.025		0.25		0.025		0.025
$CoCl_2.6H_2O$	0.025	0.025				0.025		
$Co(NO_3)_2.6H_2O$							0.05	
Na_2EDTA	37.3	37.3		37.3		74.6	37.3	37.3
$FeSO_4.7H_2O$	27.8	27.8		27.8		25.0	27.8	27.8
$MnCl_2$						3.9	0.4	
$Fe(C_4H_4O_6)_3.2H_2O$					28.0			
Vitamins and other supplements								
Inositol	100.0	100.0		100.0				100.0
Glycine	2.0	2.0	3.0	2.0				2.0
Thiamine HCl	0.1	10.0	0.1	1.0		0.3	0.3	0.5
Pyridoxine HCl	0.5		0.1	0.5		0.3	0.3	0.5
Nicotinic acid	0.5		0.5	0.5			1.25	5.0
Ca-panthothenate			1.0					

Cysteine HCl			1.0					
Riboflavin						0.3	0.05	
Biotin							0.05	0.05
Folic acid							0.3	0.5

Media Preparation

Preparation of culture media is preferred to be performed in an equipped for this purpose compartment. This compartment should beconstructed so as to maintain ease in cleaning and reducing possibility of contamination. Supplies of both tap and distilled water and gas should be provided. Appropriate systems for water sterilization or deionization are also important. Certain devices are required for better performance such as a refrigerator, freezer, hot plate, stirrer, pH meter, electric balances with different weighing ranges, heater, Bunsen burner in addition to glassware and chemicals. It is well known now that mistakes which occur in tissue culture process most frequently originate from inaccurate media preparation that is why clean glassware, high quality water, pure chemicals and careful measurement of media components should be facilitated.

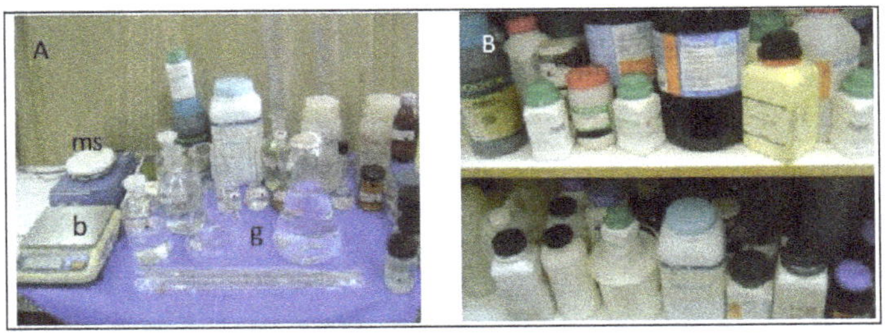

Figure: Some of components of the preparation room. A, some equipments used; ms magnetic stirrer hot plate, b electric balance, g glassware. B shelves for keeping chemicals.

A convenient method for preparation of culture media is to make concentrated stock solutions which can be immediately diluted to preferred concentration before use. Solutions of macronutrients are better to be prepared as stock solutions of 10 times the strength of the final operative medium. Stock solutions can be stored in a refrigerator at 2- 4 °C. Micronutrients stock solutions are made up at 100 times of the final concentration of the working medium. The micronutrients stock solution can also be stored in a refrigerator or a freezer until needed. Iron stock solution should be 100 times concentrated than the final working medium and stored in a refrigerator. Vitamins are prepared as either 100 or 1000 times concentrated stock solutions and stored in a freezer (-20 °C) until used if it is desired to keep them for long otherwise they can be stored in a refrigerator for 2-3 months and should be discarded thereafter. Stock solutions of growth regulators are usually prepared at 100-1000 times the final desired concentration.

Concentrations of inorganic and organic components of media are generally expressed in mass values (mg.l^{-1}, mg/l and p.p.m.). The International Association for plant

Physiology has recommended the use of mole values. Mole is an abbreviation for gram molecular weight which is the formula weight of a substance in grams. The formula weight of a substance is equal to the sum of weights of the atoms in the chemical formula of the substance. One liter of solution containing 1mole of a substance is 1 molar (1M) or 1 mol.l^{-1} solution of the substance (1 mol.l^{-1}= 103 mmol.l^{-1}= 106 µmol/l). It is routinely now to accepted to express concentrations of macronutrients and organic nutrients in the culture medium as mmol/l values, and µmol/l values for micronutrients, hormones, vitamins and other organic constituents. This was explained on the basis that mole values for all compounds have constant number of molecules per mole.

Media Selection

For the establishment of a new protocol for a specific purpose in tissue culture, a suitable medium is better formulated by testing the individual addition of a series of concentrations of a given compound to a universal basal medium such as MS, LS or B_5. The most effective variables in plant tissue culture media are growth regulators, especially auxins and cytokinins. Full strength of salts in media proved good for several species, but in some species the reduction of salts level to ½ or ¼ the full concentration gave better results in in vitro growth. Sucrose is often assumed to be the best source of carbon for in vitro culture, the levels used are from 2 to 6% and the level has to be defined for each species.

Media Sterilization

Prevention of contamination of tissue culture media is important for the whole process of plant propagation and helps to decrease the spread of plant parasites. Contamination of media could be controlled by adding antimicrobial agents, acidification or by filtration through microporous filters. To reduce possibilities of contamination, it is recommended that sterilization rooms should have the least number of openings. Media preparation and sterilization are preferred to be performed in separate compartments. Sterilization area should also have walls and floor that withstand moisture, heat and steam.

Sterilization of media is routinely achieved by autoclaving at the temperature ranging from 115° – 135 °C. Advantages of autoclaving are: the method is quick and simple, whereas disadvantages are the media pH changes and some components may decompose and so to loose their effectiveness. As example autoclaving mixtures of fructose, glucose and sucrose resulted in a drop in the agar gelling capacity and affecting pH of the culture medium through the formation of furfural derivatives due to sucrose hydrolysis.

Filtration through microporus filters (0.22- 0.45) is also used for thermolabile organic constituents such as vitamins, growth regulators and amino acids. Additives of antimicrobial agents are less commonly applied in plant tissue culture media. Limitation for their use was reported and attributed to harm imposed on plants as well.

Culture Types

'Explants' are the commonly used culture types in plant tissue culture and may consist of pieces of organs, such as leaves or roots, or may be specific cell types, such as pollen or endosperm. The efficiency of culture initiation is achieved by the explants which are derived from younger, more rapidly growing tissue (or tissue at an early stage of development). The other culture types like callus which is generated by culturing explants on appropriate medium usually with both an auxin and a cytokinin which give rise to an unorganised, growing and dividing mass of cells. The proliferation can be maintained indefinitely, provided that the callus is subcultured on to fresh medium periodically. During callus formation there is some degree of dedifferentiation (i.e. the changes that occur during development and specialization are, to some extent, reversed), both in morphology (callus is usually composed of unspecialized parenchyma cells) and metabolism. One major consequence of this dedifferentiation is that most plant cultures lose the ability to photosynthesize. This has important consequences for the culture of callus tissue, as the metabolic profile will probably not match that of the donor plant. This necessitates the addition of other components—such as vitamins and, most importantly, a carbon source—to the culture medium, in addition to the usual mineral nutrients.

Organogenesis

Organogenesis is a common method of micropropagation that involves tissue regeneration of adventitious organs or axillary buds directly or indirectly from the explants. Embryogenesis (non-zygotic) is a unique developmental pathway that is highly comparable to that of zygotic embryos and it is an important pathway for producing somaclonal variants, developing artificial seeds, and synthesizing metabolites. Since non-zygotic embryos are originated from single cell, they are favored in several regeneration systems likemicropropagation, ploidy manipulation, gene transfer, and synthetic seed production. However, tissue regeneration via organogenesis has proved to be advantageous for studying regulatory mechanisms of plant development. Because of differences in the stage of the cells in the cell cycle, the availability of or ability to transport endogenous growth regulators, and the metabolic capabilities of the cells, one could notice the specific differences in the regeneration potential of different organs and explants. The most commonly used tissue explants like the stem tip, auxiliary bud tip and root tip, as they are rich in meristematic cells; these tissues have high rates of cell division and either concentrate or produce required growth regulating substances including auxins and cytokinins. Shoot regeneration efficiency often varies between plant species and within a plant species among subspecies, varieties, cultivars, or ecotypes, therefore, tissue culture regeneration can become complicated especially when many regeneration procedures have to be developed for different genotypes within the same species. Somatic Embryogenesis and Organogenesis can be used as methods of micropropagation and sometimes they are not desirable because they may not always result in populations of identical plants. However these two methods have the beneficial use in the production of whole plants from a single cell (or a few cells). Development of organized structures can follow one of three pathways:

- Shoot regeneration, based on a unipolar structure with a shoot apical meristem.
- Root regeneration, essentially a unipolar structure with a root apical meristem.
- Somatic embryogenesis in which there is a bipolar structure.

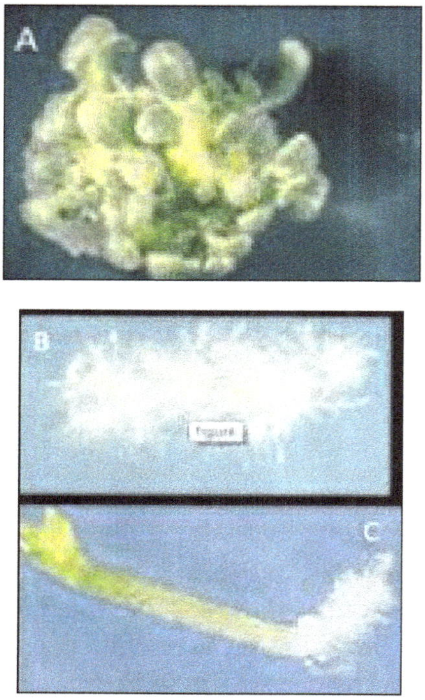

Growth in plant tissue cultures can take many forms ranging from (A) Shoot regeneration from a callus derived from a single protoplast of Hyoscyamusmuticus. (B) Root regeneration from a callus derived from a root explant of Arabidopsis thaliana. (C) A single somatic embryo removed from a callus derived from a carrot cotyledon explant.

Somatic Embryogenesis

Somatic embryogenesis is a process by which somatic cells or tissues develop into differentiated embryos. These somatic embryos can develop into whole plants without undergoing the process of sexual fertilization as done by zygotic embryos. The somatic embryogenesis can be initiated either directly from the explants or callus (mass of unorganized cells), thus regeneration of plant via somatic embryogenesis occurs by the induction of embryogenic cultures from zygotic seed, leaf or stem segment and leads to further multiplication of embryos. These mature embryos are further cultured for germination and development of plantlets, and finally transferred to soil. The plant regeneration of many of the plant species, trees and ornamental plants of different families has been achieved by Somatic embryogenesis including the some of the cactus species. There are various factors that affect the induction and development of somatic embryos in cultured cells like type of liquid medium used in culturing the explants, type and rations of plant growth. For example: highest efficiency of embryonic callus was induced by culturing nodal stem segments of rose hybrids on medium supplemented with various PGR's alone or in combination and showed highest germination rate of somatic embryos when grown on abscisic acid (ABA) alone. Thus somatic embryogenesis is not only remained as a valuable tool for genetic manipulation, but also exploited in the process of developing the plants that are resistant to various kinds of stresses.

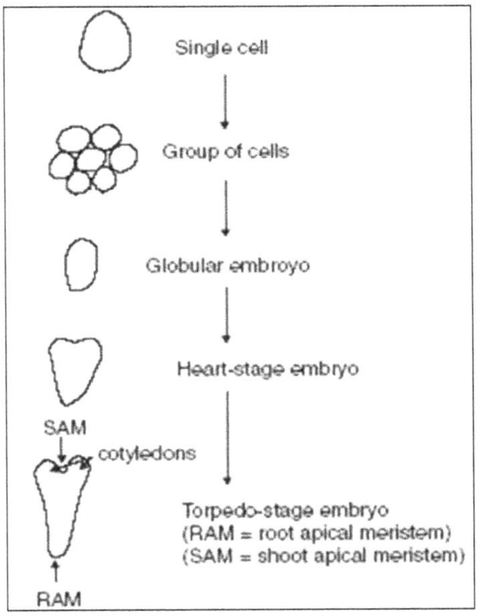

Figure: A schematic representation of the sequential stages of somatic embryo development. Somatic embryos are developed from single cells or from a small group of cells, several rounds of repeated cell divisions leads to the production of a group of cells that develop into an organised initially into 'globular-stage embryo' and switch on to heart- and torpedo stage embryos, from which plants are regenerated. Whereas zygotic embryos undergo similar development through the globular – stage embryo which is formed after the 16 cell stage and then develop into a heart and torpedo stages. Polarity is established early in embryo development. Signs of tissue differentiation become apparent at the globular stage and apical meristems are apparent in heart-stageembryos.

Organogenesis involves the regulation of cell division, cell expansion, differentiation of cell and tissue type, and patterning of the organ as a whole. Although it is highly essential to have insight into how cells are differentiating and developed to give rise to a complete organ, however, it is often directed to technical difficulties as these processes landed deep in tissues and thus are difficult to access or visualize.

An adult plant consists of highly organized tissues and organs; tissue like meristem, cortex, phloem and epidermis which possess the cells of uniform shape and specialized function and several of these tissues are organized together to form an organ, such as leaves, roots, flowers and the vascular system. The process of initiation and development of an organ is called organogenesis which is an important way to regenerate plants from the culture.The two distinct phases of Organogenesis in plant tissue culture involves firstly dedifferentiation and secondly redifferentiation. Dedifferentiation deals with the isolation of the explant tissues with an acceleration of cell division and a consequent formation of a mass of undifferentiated cells (called callus).The developmentalprocesses occur after the first callus formation is called by the name redifferentiation. In this process, the tissue called organ primordia is differentiated from a single or a group of callus cellsis generated into small meristems with cells densely filled with protoplasm and strikingly large nuclei. The Polarity (monopolar) of the longitudinal axis of the organizing growing points of the organs can be seen some time after the formation of meristem tissues and different types of specialized cells again undergo differentiation to give rise to the vascular system which is formed by connecting the new organs to their parent explants or callus mass.

As we have already discussed about the organogenesis process which relies on the production of organs either directly from an explant or callus structure, this can also be explained by the following characteristics:

- The ability of non-meristematic plant tissues to form different organs de novo.
- The formation of adventitious organs.
- The production of roots, shoots or leaves.
- These organs may arise out of pre-existing meristems or out of differentiated cells.
- This may involve a callus intermediate but often occurs without callus.

Indirect Organogenesis: Plant Growth Regulators and Differentiation

Indirect organogenesis can be written in simple forms like Explant → Callus → Meristemoid → Primordium.

Skoog and Miller have contributed tremendously to understand the mechanisms of regeneration of whole plants, or parts of plants, from cells; they observed that the direction of differentiation could be influenced by the ratio of the exogenously supplied growth regulators auxin and cytokinin and found that,a high ratio of auxin to cytokinin led to initiation of roots whereas a low ratio led to development of shoots in tobacco stem pith cultures. However for many plant species, this simple manipulation will not work, in general auxins (e.g. IAA (indoleacetic acid), NAA (a-naphthaleneacetic acid) and IBA indolebutyric acid)) will stimulate regeneration of roots, and cytokinins and kinetin will promote regeneration of shoots or embryos. Dedifferentiation and callus formation occur naturally in response to wounding and auxin and cytokinins act as a biological trigger for plant regeneration from somatic cells. With very few exceptions, auxin is essential for dedifferentiation and commonly 2,4-dichlorophenoxyacetic acid (2,4-D) is used to promote callus; cytokinin often enhances this process. Not all living cells respond to auxin and this is particularly true of mature cells of grasses. Without dedifferentiation, it is not possible to move to the next stage of totipotent expression – plant regeneration. Until the late 1980s, grasses, especially the economically important cereals, were regarded as recalcitrant in vitro. Since that time, researchers have developed methods of regeneration for most of these species. This includes work on sugar cane, sorghum, wheat and barley in several Australian laboratories.

Protoplasts capable of undergoing cell division seldom give rise directly to organized structures but first synthesize a new cell wall, then produce a callus from which shoots or embryos can later regenerate. Although auxin stimulates initial cell division in quiescent cells, continued presence of auxin can inhibit organized out-growth. In practical terms, cultures are usually transferred onto low or zero auxin media to permit or speed up shoot organogenesis. Sometimes 'removal' of auxin occurs when auxin in the medium is degraded either by the tissue itself or via chemical reactions such as photo-oxidation. Cytokinins promote out-growth of shoots but are normally kept at very low concentration when root regeneration is wanted.

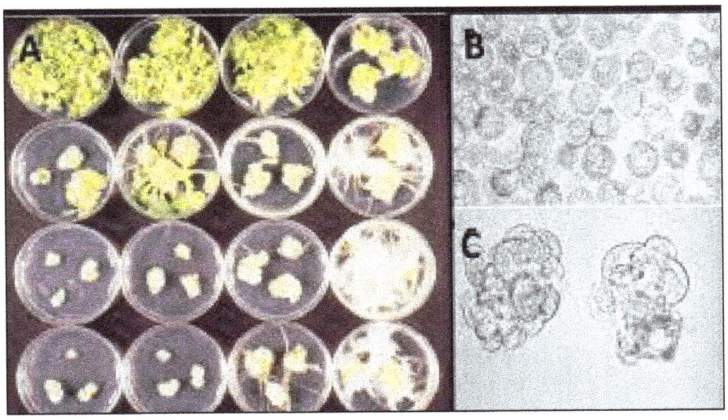

Figure shows: (A)Tobacco leaf explants cultured on media with varying concentrations of an auxin (α-naphthaleneacetic acid; NAA) and a cytokinin (6-benzylaminopurine; BAP). Concentrations of NAA are from left to right, 0, 0.01 µM, 0.1 µM, 1.0 µM; concentrations of BAP are from top to bottom 0, 0.01 µM, 0.1 µM, 1.0 µM. At low auxin to cytokinin ratios shoot development predominates, whereas at high ratios profuse root initiation occurs. At intermediate ratios, callus is often the result. Not that type of response also depends on the explant source: tobacco stem pith under the same culture conditions usually shows no development whatsoever unless some auxin is supplied (B) Protoplasts of Triticum aestivum just after isolation from leaf mesophyll cell. (C) Cell division in two protoplasts of Hyoscyamus muticus has led to formation of two cell clusters or micro-calluses.

Direct Organogenesis: The Role of Growth Regulators

Direct organogenesis is a process which bypasses the need for a callus phase. For example: formation of somatic embryos through direct embryogenesis proceeds from cells which were already embryogenically competent while they were part of the original, differentiated tissue. The formation of these pre-embryogenic cells favored only in certain conditions like wounding or application of exogenous growth regulators which allows the process of cell division and expression of embryogenesis and are tend to be more responsive than the pre embryonic cells derived from indirect organogenesis and direct organogenesis do not require the auxin 'push' to initiate division. However, cytokinin play an important role in switching between shoot development and embryogenesis; for example, in Trifoliumrepens hypocotyl epidermis, it was noticed that, BAP (a cytokinin) promotes reorientation of the plane of cell division, leading to initiation of a promeristemoid.Whereas in Abiesamabilis an analogous response occurs in cotyledon explants in which subepidermal cells develop into shoots. In case of Brassica napus anther cultures, cytokinin suppresses secondary embryoid formation and instead promotes normal leafy shoots.

Tran Thanh Van et al. studied the importance of cytokinin:auxin ratio in differentiation process, they showed that, working with thin-layer explants only three to six epidermal cells deep from floral branches of Nicotianatabacum, revealed formation of the vegetative shoot buds at the ratio of 0.1:1 auxin to cytokininandroot generation at 100:1, but a 1:1 ratio promotes floral bud initiation. Incidentally, this remains a classic example of formation of new floral meristems in vitro.

Plant Tissue Culture Techniques and Applications in Plant Improvement.	
Technique	Applications
Seed culture	Increasing efficiency of germination of seeds that are difficult to germinate in vivo. Precocious germination by application of plant growth regulators Production of clean seedlings for explants or meristem culture.
Embryo culture	Overcoming embryo abortion due to incompatibility barriers Overcoming seed dormancy and self-sterility of seeds Embryo rescue in distant (interspecific or intergeneric) hybridization where endosperm development is poor Shortening of breeding cycle.
Ovary or ovule culture	Production of haploid plants A common explant for the initiation of somatic embryogenic Cultures Overcoming abortion of embryos of wide hybrids at very early stages of development due to incompatibility barriers In vitro fertilization for the production of distant hybrids avoiding style and stigmatic incompatibility that inhibits pollen germination and pollen tube growth.
Anther and microspore culture	Production of haploid plants Production of homozygous diploid lines through chromosome doubling, thus reducing the time required to produce inbred lines. Uncovering mutations or recessive phenotypes.
In vitro pollination	Production of hybrids difficult to produce by embryo rescue.

Organ culture	Any plant organ can serve as an explant to initiate cultures.
Shoot apical meristem culture	Production of virus free germplasm Mass production of desirable genotypes Facilitation of exchange between locations (production of clean material) Cryopreservation (cold storage) or in vitro conservation of germplasm.
Somatic embryogenesis	One major path of regeneration Mass multiplication Production of artificial seeds As source material for embryogenic protoplasts Amenable to mechanization and for bioreactors.
Organogenesis	One major path of regeneration. Mass multiplication. Conservation of germplasm at either normal or sub-zero temperatures. Enhanced axillary budding Micropropagation.
Enhanced axillary budding	Micropropagation
Callus Cultures	In some instances it is necessary to go through a callus phase prior to regeneration via somatic embryogenesis or organogenesis For generation of useful somaclonal variants (genetic or epigenetic) As a source of protoplasts and suspension cultures For production of metabolites. Used in in vitro selection.
In vitro mutagenesis	Induction of polyploidy Introduction of genetic variability.
Protoplast isolation, culture and fusion	Combining genomes to produce somatic hybrids, asymmetric hybrids or cybrids. Production of organelle recombinants Transfer of cytoplasmic male sterility.
Micrografting	Overcoming graft incompatibility Rapid mass propagation of elite scions by grafting onto rootstocks that have desirable traits like resistance to soilborne pathogens and diseases. To allow survival of difficult to root shoots Development of virus free plants.
Genetic transformation	Many different explants can be used, depending on the plant species and its favored method of regeneration as well as the method of transformation. Introduction of foreign DNA to generate novel (and typically desirable) genetic combinations Used to study the function of genes.

3

Plant Biotechnology: Cell Culture and Protoplast Fusion

Plant cell culture is a collection of techniques used to grow and maintain plant cells under sterile conditions on a nutrient medium. Protoplast fusion is a physical phenomenon in which two or more protoplasts of living cells come in contact and fuse together. The diverse aspects of cell culture and protoplast fusion have been thoroughly discussed in this chapter.

Cell Culture

In spite of advances in synthetic organic chemistry, chemical synthesis of several compounds is not yet feasible due to their complex structures. The plants still contribute significantly to the bulk of the market products, such as secondary metabolites. The major limitations to the commercial use of potential metabolites is their very scarce supply from the field grown plants due to their seasonal growth, genetic, geographical and climatic variations, and insect and pathogen attack. The environmental fluxes cause alterations in type and quantity of metabolites produced. In this context, the cell culture technique is complimentary and may provide competitive metabolite production systems when compared to whole plant extraction. Since the plant cells are totipotent in nature, consequently, the cultured cells will also contain the genetic information for the production of therapeutic compounds with added potential of increasing yield by explants selection and manipulation of culture conditions. The cell cultures have several advantages over conventional isolation of metabolites from the intact plants, such as stable supply, freedom from disease and vagaries of climates, closer relationship between supply and demand, and growth of large amount of plant tissues in minimal space. Such an alternative method has considerable implication as it would reduce the pressure on natural population and, thus, may prevent the plant from becoming endangered. It may also help to steer clear of contaminated pharmaceutical raw material.

Also, studies on secondary metabolites require an in-depth understanding of biosynthetic pathways which is often difficult to conduct in whole plants because the biosynthetic activities may only be expressed in particular cell types within a specific plant organ or at a certain time of season. Cell cultures have a higher rate of metabolism than intact differentiated plants because the initiation of cell growth in culture leads to fast proliferation of cell mass and to a condensed biosynthetic cycle. As a result, secondary metabolite production can take place within a short cultivation time (about 2-4 weeks) with an added advantage of tunability. These cell cultures can be employed in scaled-up operation, for isolation of desirable compounds in bulk.

History and Evolution

Haberlandt attempted to cultivate isolated plant cells, but cell division was never observed in these cultures. In the 1930s the first *in vitro* cultures were established, and followed by the development of culture media and culture methods. Twenty-five years ago, the prospect of the use of plant cell cultures for metabolite production is not imaginable. The low yields of metabolite production in suspension cultures were the bottlenecks for commercialization. In such early efforts, plant cells in culture were treated in direct analogy to microbial systems, with little knowledge of plant cell physiology and biochemistry. In 1982, at least 30 compounds were known to accumulate in plant culture systems in concentrations equal to or higher than that of the plant. Concerning the production of natural metabolites, some *in vitro* systems exist that allow the large-scale production of economically important metabolites. A survey of the historical milestones in plant cell cultures is given by Schmauder and Doebel in 1990. Strategies to optimize growth and product formation began to develop separately during the period between 1975 and 1985. A combination of strategies for yield improvement resulted in the first commercial plant cell process.

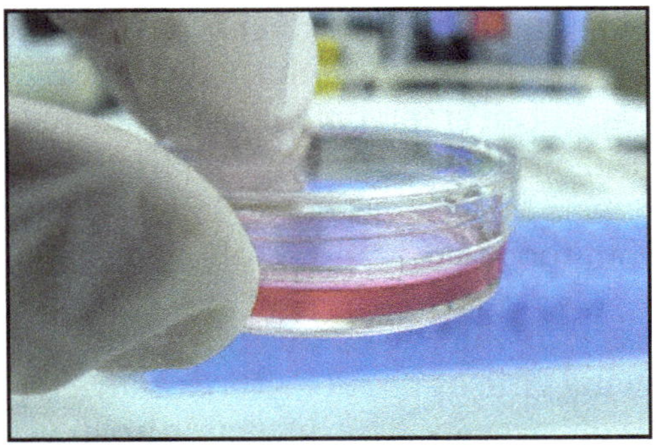

Cell culture in a special tissue culture dish.

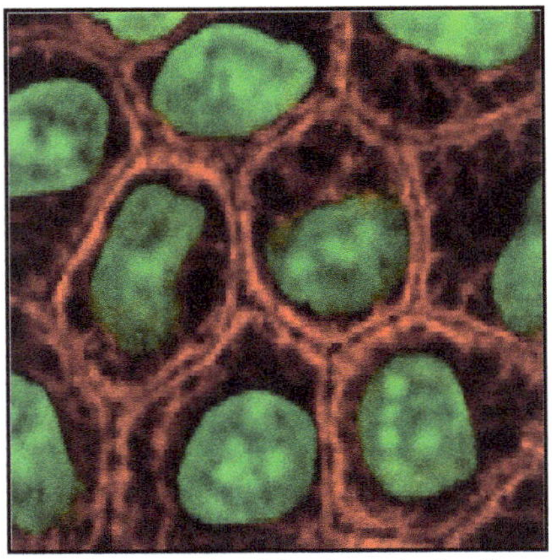

Epithelial cells in culture, stained for keratin (red) and DNA (green).

Cell culture is the process by which cells are grown under controlled conditions, generally outside of their natural environment. Cell culture conditions can vary for each cell type, but artificial environments consist of a suitable vessel with substrate or medium that supplies the essential nutrients (amino acids, carbohydrates, vitamins, minerals), growth factors, hormones, and gases (CO_2, O_2), and regulates the physio-chemical environment (pH buffer,osmotic pressure, temperature). Most cells require a surface or an artificial substrate (adherent or monolayer culture) whereas others can be grown free floating in culture medium (suspension culture).

In practice, the term "cell culture" now refers to the culturing of cells derived from multicellular eukaryotes, especially animal cells, in contrast with other types of culture that also grow cells, such as plant tissue culture, fungal culture, and microbiological culture (of microbes). The historical development and methods of cell culture are closely interrelated to those of tissue culture and organ culture. Viral culture is also related, with cells as hosts for the viruses.

The laboratory technique of maintaining live cell lines (a population of cells descended from a single cell and containing the same genetic makeup) separated from their original tissue source became more robust in the middle 20th century.

History

The 19th-century English physiologist Sydney Ringer developed salt solutions containing the chlorides of sodium, potassium, calcium and magnesium suitable for maintaining the beating of an isolated animal heart outside of the body. In 1885, Wilhelm Roux removed a portion of the medullary plate of an embryonic chicken and maintained it in a warm saline solution for several days, establishing the principle of tissue culture.Ross Granville Harrison, working at Johns Hopkins Medical School and then at Yale University, published results of his experiments from 1907 to 1910, establishing the methodology of tissue culture.

Cell culture techniques were advanced significantly in the 1940s and 1950s to support research in virology. Growing viruses in cell cultures allowed preparation of purified viruses for the manufacture of vaccines. The injectable polio vaccine developed by Jonas Salk was one of the first products mass-produced using cell culture techniques. This vaccine was made possible by the cell culture research of John Franklin Enders, Thomas Huckle Weller, and Frederick Chapman Robbins, who were awarded a Nobel Prize for their discovery of a method of growing the virus in monkey kidney cell cultures.

Concepts in Mammalian Cell Culture

Isolation of Cells

Cells can be isolated from tissues for *ex vivo* culture in several ways. Cells can be easily purified from blood; however, only the white cells are capable of growth in culture. Mononuclear cells can be released from soft tissues by enzymatic digestion with enzymes such as collagenase, trypsin, or pronase, which break down the extracellular matrix. Alternatively, pieces of tissue can be placed in growth media, and the cells that grow out are available for culture. This method is known as explant culture.

Cells that are cultured directly from a subject are known as primary cells. With the exception of some derived from tumors, most primary cell cultures have limited lifespan.

An established or immortalized cell line has acquired the ability to proliferate indefinitely either through random mutation or deliberate modification, such as artificial expression of the telomerasegene. Numerous cell lines are well established as representative of particular cell types.

Maintaining Cells in Culture

For the majority of isolated primary cells, they undergo the process of senescence and stop dividing after a certain number of population doublings while generally retaining their viability (described as the Hayflick limit).

Cells are grown and maintained at an appropriate temperature and gas mixture (typically, 37 °C, 5% CO_2 for mammalian cells) in a cell incubator. Culture conditions vary widely for each cell type, and variation of conditions for a particular cell type can result in different phenotypes.

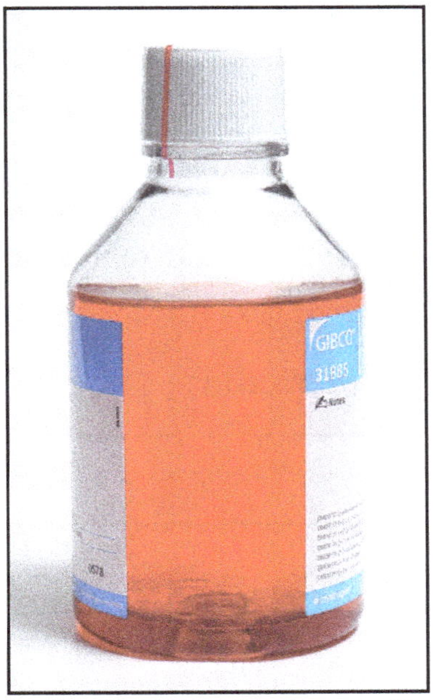

A bottle of DMEM cell culture medium.

Aside from temperature and gas mixture, the most commonly varied factor in culture systems is the cell growth medium. Recipes for growth media can vary in pH, glucose concentration, growth factors, and the presence of other nutrients. The growth factors used to supplement media are often derived from the serum of animal blood, such as fetal bovine serum (FBS), bovine calf serum, equine serum, and porcine serum. One complication of these blood-derived ingredients is the potential for contamination of the culture with viruses or prions, particularly in medical biotechnology applications. Current practice is to minimize or eliminate the use of these ingredients wherever possible and use human platelet lysate (hPL). This eliminates the worry of cross-species contamination when using FBS with human cells.

hPL has emerged as a safe and reliable alternative as a direct replacement for FBS or other animal serum. In addition, chemically defined media can be used to eliminate any serum trace (human or animal), but this cannot always be accomplished with different cell types. Alternative strategies involve sourcing the animal blood from countries with minimum BSE/TSE risk, such as The United States, Australia and New Zealand, and using purified nutrient concentrates derived from serum in place of whole animal serum for cell culture.

Plating density (number of cells per volume of culture medium) plays a critical role for some cell types. For example, a lower plating density makes granulosa cells exhibit estrogen production, while a higher plating density makes them appear as progesterone-producing theca lutein cells.

Cells can be grown either in suspension or adherent cultures. Some cells naturally live in suspension, without being attached to a surface, such as cells that exist in the bloodstream. There are also cell lines that have been modified to be able to survive in suspension cultures so they can be grown to a higher density than adherent conditions would allow. Adherent cells require a surface, such as tissue culture plastic or microcarrier, which may be coated with extracellular matrix (such as collagen and laminin) components to increase adhesion properties and provide other signals needed for growth and differentiation. Most cells derived from solid tissues are adherent. Another type of adherent culture is organotypic culture, which involves growing cells in a three-dimensional (3-D) environment as opposed to two-dimensional culture dishes. This 3D culture system is biochemically and physiologically more similar to *in vivo* tissue, but is technically challenging to maintain because of many factors (e.g. diffusion).

Components of Cell Culture Media

Component	Function
Carbon source (glucose/ glutamine)	Source of energy
Amino acid	Building blocks of protein
Vitamins	Promote cell survival and growth
Balanced salt solution	An isotonic mixture of ions to maintain optimum osmotic pressure within the cells and provide essential metal ions to act as cofactors for enzymatic reactions, cell adhesion etc.
Phenol red dye	pH indicator. The color of phenol red changes from orange/red at pH 7-7.4 to yellow at acidic (lower) pH and purple at basic (higher) pH.
Bicarbonate /HEPES buffer	It is used to maintain a balanced pH in the media

Growth Conditions

Parameter	
Temperature	37°C
CO2	5%
Humidity	95%

Cell line Cross-contamination

Cell line cross-contamination can be a problem for scientists working with cultured cells. Studies suggest anywhere from 15–20% of the time, cells used in experiments have been misidentified or contaminated with another cell line. Problems with cell line cross-contamination have even been detected in lines from the NCI-60 panel, which are used routinely for drug-screening studies. Major cell line repositories, including the American Type Culture Collection (ATCC), the European Collection of Cell Cultures (ECACC) and the German Collection of Microorganisms and Cell Cultures (DSMZ), have received cell line submissions from researchers that were misidentified by them. Such contamination poses a problem for the quality of research produced using cell culture lines, and the major repositories are now authenticating all cell line submissions. ATCC uses short tandem repeat (STR) DNA fingerprinting to authenticate its cell lines.

To address this problem of cell line cross-contamination, researchers are encouraged to authenticate their cell lines at an early passage to establish the identity of the cell line. Authentication should be repeated before freezing cell line stocks, every two months during active culturing and before any publication of research data generated using the cell lines. Many methods are used to identify cell lines, including isoenzyme analysis, human lymphocyte antigen (HLA) typing, chromosomal analysis, karyotyping, morphology and STR analysis.

One significant cell-line cross contaminant is the immortal HeLa cell line.

Other Technical Issues

As cells generally continue to divide in culture, they generally grow to fill the available area or volume. This can generate several issues:

- Nutrient depletion in the growth media.
- Changes in pH of the growth media.
- Accumulation of apoptotic/necrotic (dead) cells.
- Cell-to-cell contact can stimulate cell cycle arrest, causing cells to stop dividing, known as contact inhibition.

- Cell-to-cell contact can stimulate cellular differentiation.
- Genetic and epigenetic alterations, with a natural selection of the altered cells potentially leading to overgrowth of abnormal, culture-adapted cells with decreased differentiation and increased proliferative capacity.

Manipulation of Cultured Cells

Among the common manipulations carried out on culture cells are media changes, passaging cells, and transfecting cells. These are generally performed using tissue culture methods that rely on aseptic technique. Aseptic technique aims to avoid contamination with bacteria, yeast, or other cell lines. Manipulations are typically carried out in a biosafety hood or laminar flow cabinet to exclude contaminating micro-organisms. Antibiotics (e.g. penicillin and streptomycin) and antifungals (e.g.amphotericin B) can also be added to the growth media.

As cells undergo metabolic processes, acid is produced and the pH decreases. Often, a pH indicator is added to the medium to measure nutrient depletion.

Media Changes

In the case of adherent cultures, the media can be removed directly by aspiration, and then is replaced. Media changes in non-adherent cultures involve centrifuging the culture and resuspending the cells in fresh media.

Passaging Cells

Passaging (also known as subculture or splitting cells) involves transferring a small number of cells into a new vessel. Cells can be cultured for a longer time if they are split regularly, as it avoids the senescence associated with prolonged high cell density. Suspension cultures are easily passaged with a small amount of culture containing a few cells diluted in a larger volume of fresh media. For adherent cultures, cells first need to be detached; this is commonly done with a mixture of trypsin-EDTA; however, other enzyme mixes are now available for this purpose. A small number of detached cells can then be used to seed a new culture. Some cell cultures, such as RAW cells are mechanically scraped from the surface of their vessel with rubber scrapers.

Transfection and Transduction

Another common method for manipulating cells involves the introduction of foreign DNA by transfection. This is often performed to cause cells to express a gene of interest. More recently, the transfection of RNAi constructs have been realized as a convenient mechanism for suppressing the expression of a particular gene/protein. DNA can also be inserted into cells using viruses, in methods referred to as transduction, infection or transformation. Viruses, as parasitic agents, are well suited to introducing DNA into cells, as this is a part of their normal course of reproduction.

Established Human Cell Lines

Cell lines that originate with humans have been somewhat controversial in bioethics, as they may outlive their parent organism and later be used in the discovery of lucrative medical treatments. In the pioneering decision in this area, the Supreme Court of California held in *Moore v. Regents of the University of California* that human patients have no property rights in cell lines derived from organs removed with their consent.

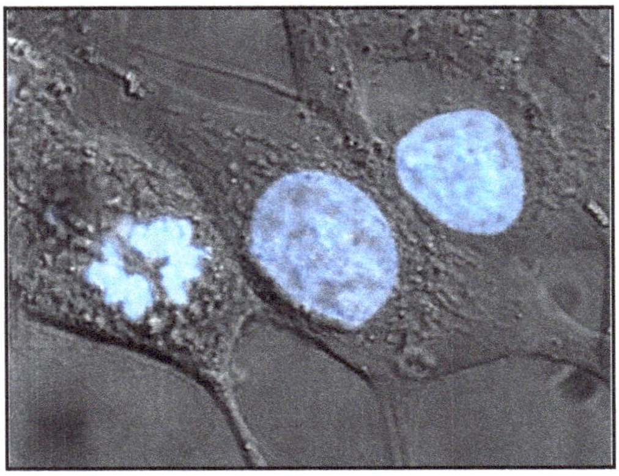

Cultured HeLa cells have been stained with Hoechst turning their nuclei blue, and are one of the earliest human cell lines descended from Henrietta Lacks, who died of cervical cancer from which these cells originated.

It is possible to fuse normal cells with an immortalised cell line. This method is used to produce monoclonal antibodies. In brief, lymphocytes isolated from the spleen (or possibly blood) of an immunised animal are combined with an immortal myeloma cell line (B cell lineage) to produce a hybridoma which has the antibody specificity of the primary lymphocyte and the immortality of the myeloma. Selective growth medium (HA or HAT) is used to select against unfused myeloma cells; primary lymphoctyes die quickly in culture and only the fused cells survive. These are screened for production of the required antibody, generally in pools to start with and then after single cloning.

Cell Strains

A cell strain is derived either from a primary culture or a cell line by the selection or cloning of cells having specific properties or characteristics which must be defined. Cell strains are cells that have been adapted to culture but, unlike cell lines, have a finite division potential. Non-immortalized cells stop dividing after 40 to 60 population doublings and, after this, they lose their ability to proliferate (a genetically determined event known as senescence).

Applications of Cell Culture

Mass culture of animal cell lines is fundamental to the manufacture of viral vaccines and other products of biotechnology. Culture of human stem cells is used to expand the number of cells and differentiate the cells into various somatic cell types for transplantation. Stem cell culture is also used to harvest the molecules and exosomes that the stem cells release for the purposes of therapeutic development.

Biological products produced by recombinant DNA (rDNA) technology in animal cell cultures include enzymes, synthetic hormones, immunobiologicals (monoclonal antibodies, interleukins, lymphokines), and anticancer agents. Although many simpler proteins can be produced using rDNA in bacterial cultures, more complex proteins that are glycosylated (carbohydrate-modified) currently must be made in animal cells. An important example of such a complex protein is the hormone erythropoietin. The cost of growing mammalian cell cultures is high, so research is underway to produce such complex proteins in insect cells or in higher plants, use of single embryonic cell and somatic embryos as a source for direct gene transfer via particle bombardment, transit gene expression and confocal microscopy observation is one of its applications. It also offers to confirm single cell origin of somatic embryos and the asymmetry of the first cell division, which starts the process.

Cell culture is also a key technique for cellular agriculture, which aims to provide both new products and new ways of producing existing agricultural products like milk, (cultured) meat, fragrances, and rhino horn from cells and microorganisms. It is therefore considered one means of achieving animal-free agriculture.

Cell Culture in Two Dimensions

Research in tissue engineering, stem cells and molecular biology primarily involves cultures of cells on flat plastic dishes. This technique is known as two-dimensional (2D) cell culture, and was first developed by Wilhelm Roux who, in 1885, removed a portion of the medullary plate of an embryonic chicken and maintained it in warm saline for several days on a flat glass plate. From the advance of polymer technology arose today's standard plastic dish for 2D cell culture, commonly known as the Petri dish. Julius Richard Petri, a German bacteriologist, is generally credited with this invention while working as an assistant to Robert Koch. Various researchers today also utilize culturing laboratory flasks, conicals, and even disposable bags like those used in single-use bioreactors.

Aside from Petri dishes, scientists have long been growing cells within biologically derived matrices such as collagen or fibrin, and more recently, on synthetic hydrogels such as polyacrylamide or PEG. They do this in order to elicit phenotypes that are not expressed on conventionally rigid substrates. There is growing interest in controlling matrix stiffness, a concept that has led to discoveries in fields such as:

- Stem cell self-renewal.
- Lineage specification.
- Cancer cell phenotype.
- Fibrosis.
- Hepatocyte function.
- Mechanosensing.

Cell Culture in Three Dimensions

Cell culture in three dimensions has been touted as "Biology's New Dimension". At present, the practice of cell culture remains based on varying combinations of single or multiple cell structures in 2D. Currently, there is an increase in use of 3D cell cultures in research areas including drug discovery, cancer biology, regenerative medicine and basic life science research. There are a variety of platforms used to facilitate the growth of three-dimensional cellular structures including scaffold systems such as hydrogel matrices and solid scaffolds, and scaffold-free systems such as low-adhesion plates, nanoparticle facilitated magnetic levitation, and hanging drop plates.

3D Cell Culture in Hydrogels

As the natural extracellular matrix (ECM) is important in the survival, proliferation, differentiation and migration of cells, different hydrogel culture matrices mimicking natural ECM structure are seen as potential approaches to in vivo –like cell culturing. Hydrogels are composed of interconnected pores with high water retention, which enables efficient transport of substances such as nutrients and gases. Several different types of hydrogels from natural and synthetic materials are available for 3D cell culture, including animal ECM extract hydrogels, protein hydrogels, peptide hydrogels, polymer hydrogels, and wood-based nanocellulose hydrogel.

3D Cell Culturing by Magnetic Levitation

The 3D Cell Culturing by Magnetic Levitation method (MLM) is the application of growing 3D tissue by inducing cells treated with magnetic nanoparticle assemblies in spatially varying magnetic fields using neodymium magnetic drivers and promoting cell to cell interactions by levitating the cells up to the air/liquid interface of a standard petri dish. The magnetic nanoparticle assemblies consist of magnetic iron oxide nanoparticles, gold nanoparticles, and the polymer polylysine. 3D cell culturing is scalable, with the capability for culturing 500 cells to millions of cells or from single dish to high-throughput low volume systems.

Tissue Culture and Engineering

Cell culture is a fundamental component of tissue culture and tissue engineering, as it establishes the basics of growing and maintaining cells *in vitro*. The major application of human cell culture is in stem cell industry, where mesenchymal stem cells can be cultured and cryopreserved for future use. Tissue engineering potentially offers dramatic improvements in low cost medical care for hundreds of thousands of patients annually.

Vaccines

Vaccines for polio, measles, mumps, rubella, and chickenpox are currently made in cell cultures. Due to the H5N1pandemic threat, research into using cell culture for influenza vaccines is being funded by the United States government. Novel ideas in the field include recombinant DNA-based vaccines, such as one made using human adenovirus (a common cold virus) as a vector, and novel adjuvants.

Culture of Non-mammalian Cells

Plant Cell Culture Methods

Plant cell cultures are typically grown as cell suspension cultures in a liquid medium or as callus cultures on a solid medium. The culturing of undifferentiated plant cells and calli requires the proper balance of the plant growth hormones auxin and cytokinin.

Insect Cell Culture

Cells derived from Drosophila melanogaster (most prominently, Schneider 2 cells) can be used for experiments which may be hard to do on live flies or larvae, such as biochemical studies or studies using siRNA. Cell lines derived from the army worm *Spodoptera frugiperda*, including Sf9 and Sf21, and from the cabbage looper *Trichoplusia ni*, High Five cells, are commonly used for expression of recombinant proteins using baculovirus.

Bacterial and Yeast Culture Methods

For bacteria and yeasts, small quantities of cells are usually grown on a solid support that contains nutrients embedded in it, usually a gel such as agar, while large-scale cultures are grown with the cells suspended in a nutrient broth.

Viral Culture Methods

The culture of viruses requires the culture of cells of mammalian, plant, fungal or bacterial origin as hosts for the growth and replication of the virus. Whole wild type viruses, recombinant viruses or viral products may be generated in cell types other than their natural hosts under the right conditions. Depending on the species of the virus, infection and viral replication may result in host cell lysis and formation of a viral plaque.

Common Cell Lines

Human cell lines:

- DU145 (prostate cancer).
- H295R (adrenocortical cancer).
- HeLa (cervical cancer).
- KBM-7 (chronic myelogenous leukemia).
- LNCaP (prostate cancer).
- MCF-7 (breast cancer).

- MDA-MB-468 (breast cancer).
- PC3 (prostate cancer).
- SaOS-2 (bone cancer).
- SH-SY5Y (neuroblastoma, cloned from a myeloma).
- T47D (breast cancer).
- THP-1 (acute myeloidleukemia).
- U87 (glioblastoma).
- National Cancer Institute's 60 cancer cell line panel (NCI60).

Primate cell lines:

- Vero (African green monkey *Chlorocebus* kidney epithelial cell line).

Mouse cell lines:

- MC3T3 (embryonic calvarium).

Rat tumor cell lines:

- GH3 (pituitary tumor).
- PC12 (pheochromocytoma).

Plant cell lines:

- Tobacco BY-2 cells (kept as cell suspension culture, they are model system of plant cell).

Other species cell lines:

- Dog MDCK kidney epithelial.
- Xenopus A6 kidney epithelial.
- ZebrafishAB9.

Immortalised Cell Line

An immortalised cell line is a population of cells from a multicellularorganism which would normally not proliferate indefinitely but, due to mutation, have evaded normal cellular senescence and instead can keep undergoing division. The cells can therefore be grown for prolonged periods *in vitro*. The mutations required for immortality can occur naturally or be intentionally induced for experimental purposes. Immortal cell lines are a very important tool for research into the biochemistry and cell biology of multicellular organisms. Immortalised cell lines have also found uses in biotechnology.

An immortalised cell line should not be confused with stem cells, which can also divide indefinitely, but form a normal part of the development of a multicellular organism.

Relation to Natural Biology and Pathology

There are various immortal cell lines. Some of them are normal cell lines - e.g. derived from stem cells. Other immortalised cell lines are the *in vitro* equivalent of cancerous cells. Cancer occurs when a somatic cell which normally cannot divide undergoes mutations which cause de-regulation of the normal cell cycle controls leading to uncontrolled proliferation. Immortalised cell lines have undergone similar mutations allowing a cell type which would normally not be able to divide to be proliferated *in vitro*. The origins of some immortal cell lines, for example HeLa human cells, are from naturally occurring cancers.

Role and Uses

Immortalised cell lines are widely used as a simple model for more complex biological systems, for example for the analysis of the biochemistry and cell biology of mammalian (including human) cells. The main advantage of using an immortal cell line for research is its immortality; the cells can be grown indefinitely in culture. This simplifies analysis of the biology of cells which may otherwise have a limited lifetime.

Immortalised cell lines can also be cloned giving rise to a clonal population which can, in turn, be propagated indefinitely. This allows an analysis to be repeated many times on genetically identical cells which is desirable for repeatable scientific experiments. The alternative, performing an analysis on primary cells from multiple tissue donors, does not have this advantage.

Immortalised cell lines find use in biotechnology where they are a cost-effective way of growing cells similar to those found in a multicellular organism *in vitro*. The cells are used for a wide variety of purposes, from testing toxicity of compounds or drugs to production of eukaryotic proteins.

Limitations

Changes from Nonimmortal Origins

While immortalised cell lines often originate from a well-known tissue type they have undergone significant mutations to become immortal. This can alter the biology of the cell and must be taken into consideration in any analysis. Further, cell lines can change genetically over multiple passages, leading to phenotypic differences among isolates and potentially different experimental results depending on when and with what strain isolate an experiment is conducted with.

Contamination with Other Cells

Many cell lines that are widely used for biomedical research have been contaminated and overgrown by other, more aggressive cells. For example, supposed thyroid lines were actually melanoma cells, supposed prostate tissue was actually bladder cancer, and supposed normal uterine cultures were actually breast cancer.

Methods of Generation

There are several methods for generating immortalised cell lines:

- Isolation from a naturally occurring cancer. This is the original method for generating an immortalised cell line. Major examples include human HeLa cells that were obtained from a cervical cancer, mouse Raw 264.7 cells that were subjected to mutagenesis and then selected for cells which are able to undergo division.
- Introduction of a viral gene that partially deregulates the cell cycle (e.g., the adenovirus type 5 E1 gene was used to immortalize the HEK 293 cell line).
- Artificial expression of key proteins required for immortality, for example telomerase which prevents degradation of chromosome ends during DNA replication in eukaryotes
- Hybridoma technology, specifically used for the generation of immortalised antibody-producing B cell lines, where an antibody-producing B cell is fused with a myeloma (B cell cancer) cell.

Examples

There are several examples of immortalised cell lines, each with different properties. Most immortalised cell lines are classified by the cell type they originated from or are most similar to biologically:

- 3T3 cells – a mouse fibroblast cell line derived from a spontaneous mutation in cultured mouse embryo tissue.
- A549 cells – derived from a cancer patient lung tumor.
- F11 cells – a line of neurons from the dorsal root ganglia of rats.
- HeLa cells – a widely used human cell line isolated from cervical cancer patient, Henrietta Lacks.
- HEK 293 cells – derived from human fetal cells.
- Jurkat cells – a human T lymphocyte cell line isolated from a case of leukemia.
- Vero cells – a monkey kidney cell line that arose by spontaneous immortalisation.

Initiation and Establishment of Cell Suspension Cultures

Callus Cultures

When an organ of a plant is damaged a wound repair response is induced to bring about the repair of the damaged portion. This response is associated with the induction of division in the undamaged cells adjacent to the lesion, thus sealing of the wound. If, however, wounding is followed by the aseptic culture of the damaged region on a defined medium, the initial cell division response can be stimulated and induced to continue indefinitely through the exogenous influence of the chemical constitution of the culture medium. The result is a continually-dividing mass of cells without any significant differentiation and organization and this proliferated mass of cell aggregate is called callus. The first step to establish cell suspension cultures is to raise callus from any explants of the plant. To maximize the production of a particular compound, it is desirable to initiate the callus from the plant part that is known to be high producer.

Calli are generally grown on medium solidified with gelling agents like, agar, gelrite, agarose, in Petri-dishes, glass test-tubes or extra-wide necked Erlenmeyer flasks. In morphological terms it can vary extensively, ranging from being very hard/compact and green or light green in color, where the cells have extensive and strong cell to cell contact, to being 'friable' where the callus consists of small, disintegrating aggregates of poorly-associated cells and has brownish or creamy appearance. Friable callus is most demanded since it shows fast and uniform growth of cells and is highly suitable to initiate cell suspension cultures. Callus morphology is explants and species dependent but can be altered by the modification of plant growth regulators in the medium.

The callus cultures shows inherent degree of heterogeneity and this may be due to their size and nature, unidirectional supply of nutrient medium (below the callus) and gases and light (predominantly from above). The heterogeneity may be disadvantageous in uniform production of cell biomass but may be useful in the developmental responses of the callus like, shoot regeneration.

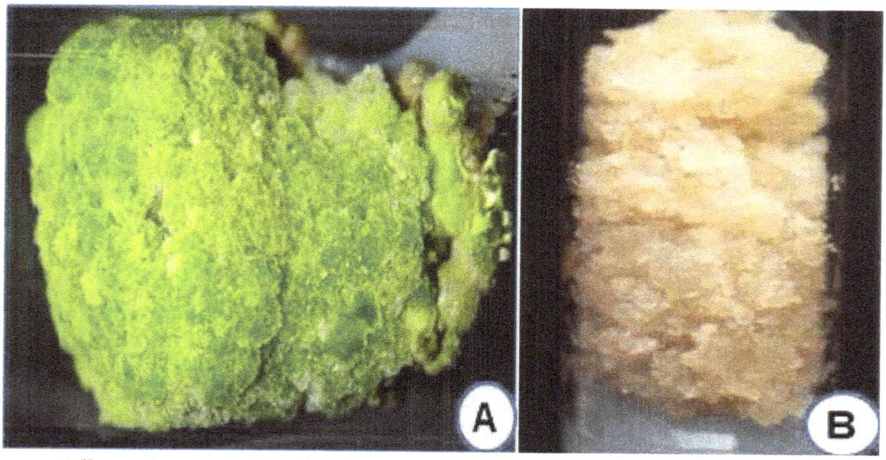

Callus cultures- **A.** Hard and compact callus; **B.** Friable and brown callus.

Cell Suspension Culture

Characteristics: A suspension culture is developed by transferring the relatively friable portion of a callus as in figure, into liquid medium and is maintained under suitable conditions of aeration, agitation, light, temperature and other physical parameters. The increased cell dissociation means increased culture uniformity. Plant cells are significantly larger and slower growing cells than most microbial organisms. They mostly resemble to parenchymatous cells in having relatively large vacuoles, a thin layer of cytoplasm and thin, rounded cell walls. The species/genotypes and medium composition used can influence *in vitro* cell morphology and different cell types with different morphological/ physiological properties can co-exist within a single culture.

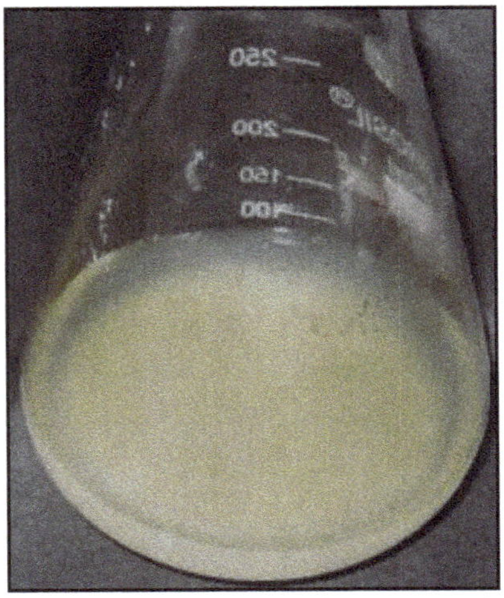
Cells in liquid medium showing fine suspension of cells.

Cell growth: The most commonly used cell suspensions are of the closed (or batch) type where the cells are grown in fixed volume of liquid medium and which are routinely maintained through the transfer of a portion (ca 10%) of a fully-grown culture to fresh medium at regular intervals. The growth curve of a cell suspension culture has a characteristic shape consisting of four essential stages- an initial lag phase, an exponential phase, stationary phase and death phase. The duration of each phase is dependent on the species or genotype selected, explant used, culture medium and subculture regime. The lag phase is shortened when relatively large inocula are used although paradoxically, growth terminates earlier and overall biomass production is reduced.

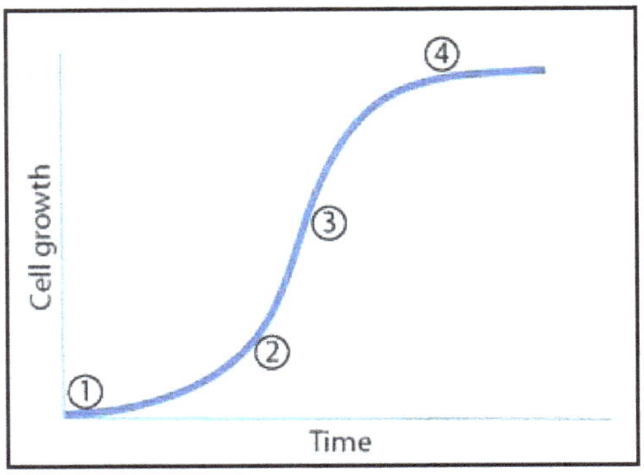

Growth curve for plant cell suspension grown in closed system. The four different growth phases are labeled: (1) Lag phase, (2) Exponential phase, (3) Linear phase, (4) Stationary phase.

Aggregation: Due to larger size of a plant cell, it is capable of withstanding tensile strain but is sensitive to shear stress. Aggregation is common, largely due to failure of the cells to separate after division. The secretion of extracellular polysaccharides, particularly in the later stages of growth, may further contribute to increased adhesion. This tendency of plant cells to grow in clumps results in sedimentation, insufficient mixing and diffusion-limited biochemical reaction. Even the fine suspension culture consists of micro-to sub-macroscopic colonies made up of around 5-200 cells and such degree of cell aggregation is acceptable. Cultures consisting of larger aggregates like, 0.5-1.0 mm in diameter, are more readily attainable, grow perfectly well and depending on the aim of the research are often sufficient to meet all requirements. This so called cell-cell contact is desirable for the biosynthesis of many secondary metabolites by the plant cells in suspension cultures. Therefore, controlled aggregation of plant cells may be of interest from process engineering point of view.

Oxygen and aeration effect: Oxygen requirements of plant cells are comparatively lower than that of microbial cells due to their low growth rates. In some cases high oxygen concentration is even toxic to the cell's metabolic activities and may strip nutrients such as carbon dioxide from the culture broth. Carbon dioxide is often considered as essential nutrient in the culture of plant cells and has a positive effect on cell growth. Moderate shaking speed like, 90-120 rpm is ideal for standard aeration. As the plant cells are shear sensitive and the immediate effects of high agitation are the cell damage, reduction in cell viability, release of intracellular compounds while low agitation (<90rpm) results in cell aggregation.

Plant Cell Cultures Vs. Microbial Cultures

Although basic equipment-and process-related requirements are same for both plant and microbial cell suspension cultures but some of the features suitable for microbial cells cannot be used for plant cells due to prominent differences in the nature and growth pattern of the two types of cells:

1. Plant cells are sensitive to shear stresses because of the relatively inflexible cell wall and their large size (50-100 μm) compared to microbial cells (2-10 μm).
2. Plant cell cultures show relatively long growth cycles. Typical specific growth rates (increase in cell mass per unit time) may range from $0.12d^{-1}$ to $0.05d^{-1}$; thus, the typical doubling time of plant cell cultures is measured in days, as compared to hours for microbial systems.
3. In plant cells, the vacuole is usually the site of product accumulation and extracellular product secretion is rare. While in microbial cells product accumulation often extracellular.
4. The plant cells mostly grow as aggregates while microbial cells grow as single cells.

5. The plant cells in suspension cultures often undergo spontaneous genetic variation in terms of accumulation of secondary metabolites, which leads to heterogeneous population of cells in a suspension culture. Compared to this microbial cells are genetically stable.

Experiment

Aim: To raise cell suspension culture of *Lantana camara*.

Equipments: Autoclave, pH-meter, Magnetic stirrers, Magnetic beads, Weighing balance, Laminar-air-flow, Microsopes, Incubator shaker.

Materials Required : Salts and vitamins of Murashige and Skoog's (MS; 1962), sucrose, agar, conical flasks, measuring cylinders and beakers of various sizes. Reagent glass bottles for storage, spatula, tissue rolls, distilled water. Cotton plugs, aluminium foils, muslin cloth, scissor, media stocks, 1N NaOH, 1N HCl, myo-inositol. Autoclavable polybags, rubber bands. Borosil glass test-tubes (150mm x 25mm without rim) and Erlenmeyer flasks. Black markers, micropipette, micropipette-tips, test-tube stands, autoclavable baskets, plastic trays, cork borer, Leaves from healthy *L. camara* plants bearing pink-yellow colored flowers.

Plant Growth regulators (Sigma):

Benzyl amino purine (BAP), Naphthalene acetic acid (NAA) , 2-4-dichlorophenoxy-acetic acid.

Methods

Initiation and establishment of leaf-disc cultures:

Leaves were washed with 1% Tween-20 (Merck, India) for 15 min, followed by three rinses in sterile distilled water (SDW). Thereafter, the remaining steps were carried out inside the laminar-air-flow cabinet (Saveer Biotech, India). Leaves were surface sterilized with 0.1% mercuric chloride solution ($HgCl_2$) for 10 min and rinsed thrice with SDW. Leaf-disc explants were prepared by punching the sterilized leaves with 5 mm sized cork-borer before being cultured with the adaxial side in contact with the media.

Establishment of callus cultures:

Leaf disc explants were incubated on MS (with 3% sucrose) basal medium consisting of macro and microsalts, vitamins, iron, 100 mgl^{-1} myoinositol and solidified with 0.8% agar (HiMedia, India). The medium was enriched with combination of cytokinin and auxins, BAP (5μM) + NAA (1 μM) + 2,4-D (1 μM) for induction and multiplication of callus from leaf-disc explants. After adjusting the pH to 5.8, 20 ml of the medium was dispensed into each 150 x 25 mm Borosil rimless glass tubes. The culture tubes were plugged with nonabsorbent cotton wrapped in cheesecloth and autoclaved at 121°C at

15 psi for 15 min. All the cultures were maintained in diffuse light (1000-1600 lux) and 16 h photoperiod at 25 ± 2°C and 50-60% relative humidity. Observations were recorded at weekly intervals. After callus induction, the biomass was multiplied constantly by inoculating 0.2 g of calli onto fresh medium at every 4-week intervals.

Establishment of cell suspension cultures:

After 7-8 passages of callus subcultures, healthy, green, friable and soft calli maintained on responding semi-solid medium were utilized to establish suspension cultures. Cultures were initiated in Erlenmeyer flasks of 250 ml capacity, containing 50 ml of liquid medium and inoculated with 0.2 g of fresh calli. The cultures were incubated in an orbital shaker under shaking conditions at 120 rpm at 25°C ± 2°C and maintained in diffuse light (1000-1600 lux) with 16 h photoperiod or in continuous darkness. The cell biomass was subcultured regularly into fresh medium at every three weeks.

Batch kinetics of cell suspension cultures:

a. Batch kinetics studies:

To determine the specific growth rate, cells were harvested from liquid medium at an interval of two days, washed and dried. The pH and conductivity of the suspension cultures were monitored after every two days. Phosphate was estimated by the standard calibration curve made from dihydrogen sodium phosphate (NaH_2PO_4); to 0.5 ml of standard or sample solution, 4 ml of reagent [Acetone (CH_3COCH_3), Sulphuric acid (H_2SO_4)2.5 M and Ammonium molybdate tetrahydrate ($(NH_4)_2MoO_4.4H_2O$) 10 mM, mixed in the ratio of 2:1:1] was added. After mixing the solutions thoroughly, 0.4 ml of 1M citric acid was added and absorbance was taken at 355 nm. Similarly, for nitrate estimation, standard curve was made from 0.01N stock solution of Potassium nitrate (KNO_3), preserved in chloroform. After acidification of samples with hydrochloric acid, absorbance was recorded at 275 nm in a UV visible spectrophotometer. All measurements and results are average readings obtained from three flasks.

b. Agitation speed and cell viability:

Effect of agitation speeds was evaluated on fresh and dry weight of cells and their viability, at the end of each passage. Callus cells weighing approximately 0.2 g were harvested at the end of growth period and re-inoculated in 50 ml of fresh medium of the same composition. The cultures were incubated in shaking conditions at 60, 120 and 240 rpm, under darkness, for a period of three weeks and their fresh and dry weights were recorded. The viability of cells under each condition was checked with 1% fluorescein diacetate (FDA) solution. FDA is a cell permeant dye. Within the cells, the molecule is cleaved by esterase activity to fluorescein which is unable to pass through the cell membrane of live cells while it leaches out from the dead cells. Hence, only the live, intact cells take up the stain and fluoresce green.

Results

Analysis of Callus Cultures

Establishing cell and tissue cultures of *Lantana* is a difficult task to accomplish because of the interference posed by phenolic compounds. The best treatment for callusing, in terms of number of explants showing callusing and the degree of callusing, was the combination of MS + 2,4-D (1 μM) + NAA (1 μM) + BAP (5 μM). On the responding medium, 100% of the explants callused, and the callus growth was profuse after the first subculture. On this medium, the leaf-disc explants first turned brown but after a week, bright-green, hard, compact calli started developing from the margins of the leaf-disc. These compact calli were dissected out and subcultured on the fresh medium of the same growth regulator composition. The rate and degree of callus proliferation increased with the subsequent subcultures, the nature of callus did not improve substantially. The callus was friable and soft but remained deep brown in nature. Until 10 th subculture, the cells in the callus were a mixture of green and brown cells. It took about 26 weeks of regular subculturing, at 4-week intervals, to obtain profusely growing fresh, friable, granulated and cream callus.

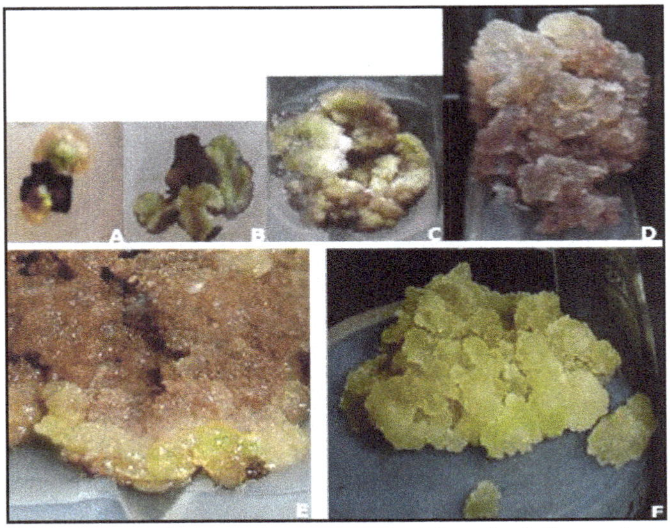

Callus culture of *Lantana camara* raised from leaf-disc explants.

Analysis of Cell Suspension Cultures Raised in Shake Flasks

Kinetics of cell growth and nutrient uptake:

The specific growth rate (μ) of the suspended cells was found to be 0.1072 d^{-1}. It was observed that the cultures remained in the lag phase till the 2nd day. Biomass increased till the 10th day following which the stationary phase started. The pH of the medium underwent variation during different stages of culture. It was observed that after showing a slight decrease in the value, the pH dropped sharply between 4-6 days, which dropped further after 10 days resulting in poor growth and stationary phase of cells. The drop of pH may be attributed to preferential uptake of NH_4^+ ions which in turn resulted in decreased pH due to liberation of H^+ ions; pH tends to increase if NO_3^- is utilized faster than NH_4^+. Uptake of nitrate was observed at slower rate. It was present in the culture medium till the last days of cultivation (16th day). The concomitant synthesis of acid triterpenes in the medium was found to be growth associated and showed an increase with the increase in biomass. Conductivity was shown to have an inverse relationship with growth. Among the major inorganic nutrients, it was invariably observed that phosphate was almost completely consumed by the 10th day of culture. Its utilization was very fast in the initial days than in the later stages of growth. Hence, it may be concluded that complete utilization of phosphate and a marked drop in the pH of the culture medium resulted in the onset of stationary phase in cell suspension cultures.

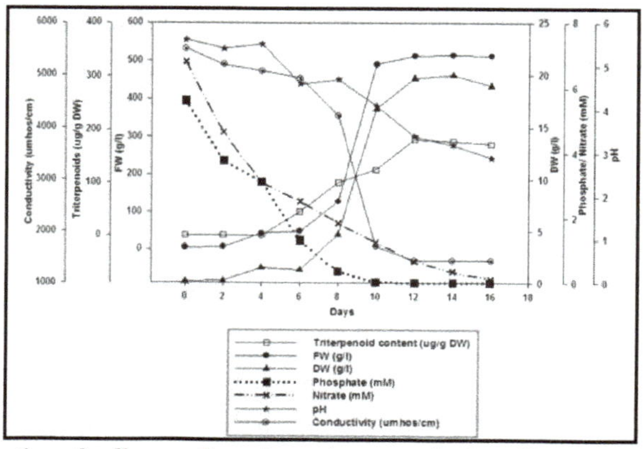

Kinetics of cell growth and nutrient uptake in cell suspension cultures of *L. camara*.

Effect of agitation speed on cell survival and viability:

Speed of agitation directly affected the growth and viability of cells in suspension cultures due to aeration and shearing effect. The maximum fresh weight and maximum viability was observed at 120 rpm. At lower agitation (60 rpm), the cells died due to aggregation and clumping; only the cells at the outermost layer of the aggregate were alive and fluorescent green when stained with fluorescein diacetate (FDA). At higher speed (240 rpm), the cells died due to rupturing.

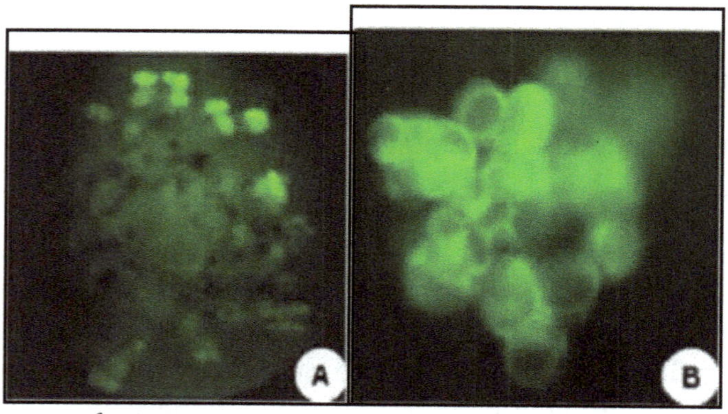

60 mm 120 mm

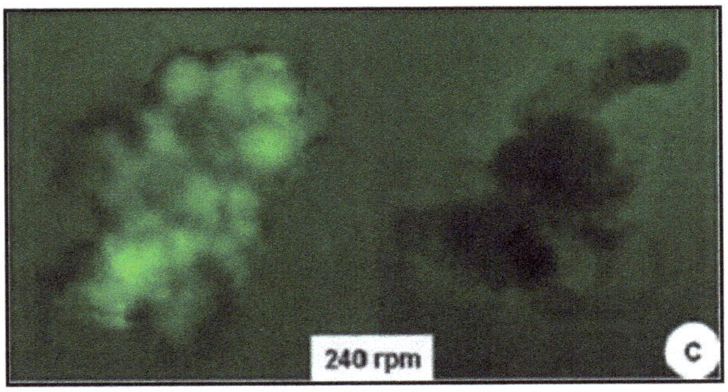

Cells stained with 1% fluorescein diacetate solution . A. cellular clump at 60 rpm, showing unstained dead cells in the centre of cell aggregate and live cells fluoresce green at the periphery , B. The cultures maintained at 120 rpm in the cell suspension, showing aggregates of live and healthy, fluorescent green stained cells , C. Same at 240 rpm, showing dead clumps (dark bodies) and sheared cells.

3D Cell Culturing by Magnetic Levitation

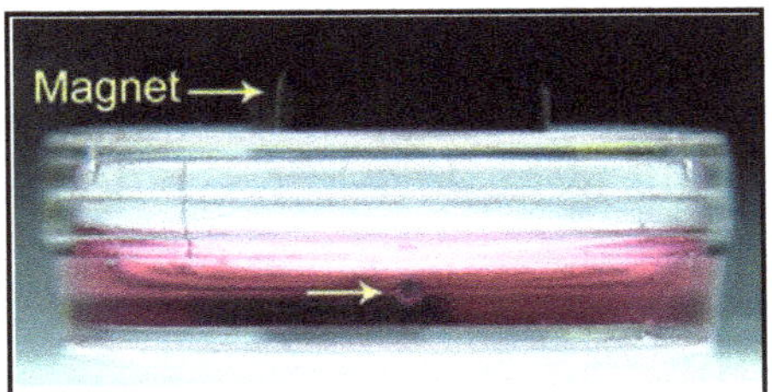

3D cell culture grown with magnetic levitation. Human Glioblastoma (hGBM) cells (indicated by lower arrow) treated with magnetic nanoparticles were held at the air-medium interface by the magnetic field created by the magnet attached to the top of the 1st tissue culture plate. This image was taken after 48 hr of culturing

3D cell culture by the magnetic levitation method (MLM) is the application of growing 3D tissue by inducing cells treated with magnetic nanoparticle assemblies in spatially varying magnetic fields using neodymium magnetic drivers and promoting cell to cell interactions by levitating the cells up to the air/liquid interface of a standard petri dish. The magnetic nanoparticle assemblies consist of magnetic iron oxide nanoparticles, gold nanoparticles, and the polymer polylysine. 3D cell culturing is scalable, with the capability for culturing 500 cells to millions of cells or from single dish to high-throughput low volume systems. Once magnetized cultures are generated, they can also be use as the building block material, or the "ink", for the magnetic 3D bioprinting process.

Overview

Standard monolayer cell culturing on tissue culture plastic has notably improved our understanding of basic cell biology, but it does not replicate the complex 3D architecture of in vivo tissue, and it can significantly modify cell properties. This often compromises experiments in basic life science, leads to misleading drug-screening results on efficacy and toxicity, and produces cells that may lack the characteristics needed for developing tissue regeneration therapies.

The future of cell culturing for fundamental studies and biomedical applications lies in the creation of multicellular structure and organization in three-dimensions. Many schemes for 3D culturing are being developed or marketed, such as bio-reactors or protein-based gel environments.

A 3D cell culturing system known as the Bio-Assembler™ uses biocompatible polymer-basedreagents to deliver magnetic nanoparticles to individual cells so that an applied magnetic driver can levitate cells off the bottom of the cell culture dish and rapidly bring cells together near the air-liquid interface. This initiates cell-cell interactions in the absence of any artificial surface or matrix. Magnetic fields are designed to rapidly form 3D multicellular structures in as little as a few hours, including expression of extracellular matrix proteins. The morphology, protein expression, and response to exogenous agents of resulting tissue show great similarity to in vivo results.

History

3D cell culturing by magnetic levitation method (MLM) was developed from collaboration between scientists at Rice University and University of Texas MD Anderson Cancer Center in 2008. Since then, this technology has been licensed and commercialized by Nano 3D Biosciences.

The Magnetic Levitation Process

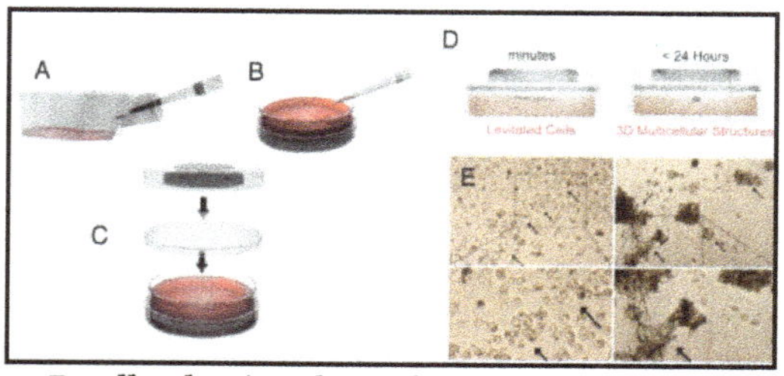

Above is a picture showing 3D cell culturing through magnetic levitation with the Bio-Assembler cell culturing system. (A) A magnetic iron oxide nanoparticle assembly known as Nanoshuttle is added and dispersed over cells and the mixture is incubated. (B) After incubation with Nanoshuttle, cells are detached and transferred to a petri dish. (C) A magnetic drive is then placed on top of a petri dish top. (D) The magnetic field causes cells to rise to the air–medium interface. (E) Human umbilical vein endothelial cells (HUVEC) levitated for 60 minutes (left images) and 4 hours (right images) (Scale bar, 50 μm). The onset of cell-cell interaction takes place as soon as cells levitate, and 3D structures start to form. At 1 hour, the cells are still relatively dispersed, but they are already showing some signs of stretching. Formation of 3D structures is visible after 4 hours of levitation (arrows).

Protein Expression

Protein expression in levitated cultures shows striking similarity to in vivo patterns. N-cadherin expression in levitated human glioblastoma cells was identical to the expression seen in human tumor xenografts grown in immunodeficient mice, while standard 2D culture showed much weaker expression that did not match xenograft distribution as shown in the picture below. The transmembrane protein N-cadherin is often used as an indicator of in-vivo-like tissue assembly in 3D culturing.

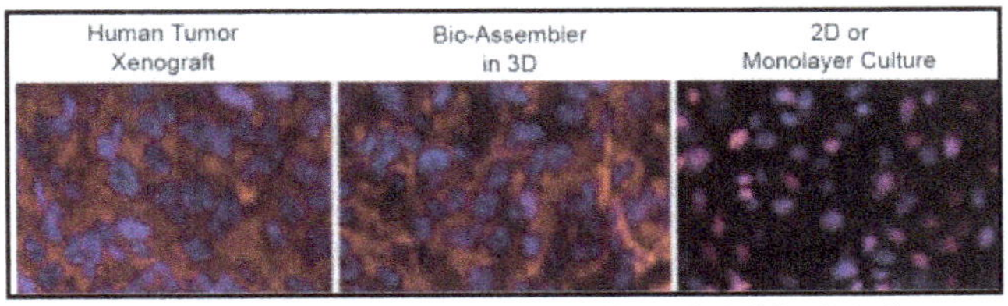

In the picture above, distribution of N-cadherin (red) and nuclei (blue) in human brain cancer mouse xenograft (left, human brain cancer cells grown in a mouse brain), brain cancer cells cultured by 3D magnetic levitation for 48 h. (middle), and cells cultured on a glass slide cover slip (2D, right). The 2D system shows N-cadherin in the cytoplasm and nucleus and notably absent from the membrane, while in the levitated culture and mouse, N-cadherin is clearly concentrated in the membrane, and also present in cytoplasm and cell junctions.

Applications

Co-culturing, Magnetic Manipulation, and Invasion Assays

One of the challenges in generating in vivo like cultures or tissue in vitro is the difficulty in co-culturing different cell types. Because of the ability of 3D cell culturing by magnetic levitation to bring cells together, co-culturing different cell types is possible. Co-culturing of different cell types can be achieved at the onset of levitation, by mixing different cell types in before levitation or by magnetically guiding 3D cultures in an invasion assay format.

The unique ability to manipulate cells and shape tissue magnetically offers new possibilities for controlled co-culturing and invasion assays. Co-culturing in a realistic tissue architecture is critical for accurately modeling in vivo conditions, such as for increasing the accuracy of cellular assays as shown in the figure below.

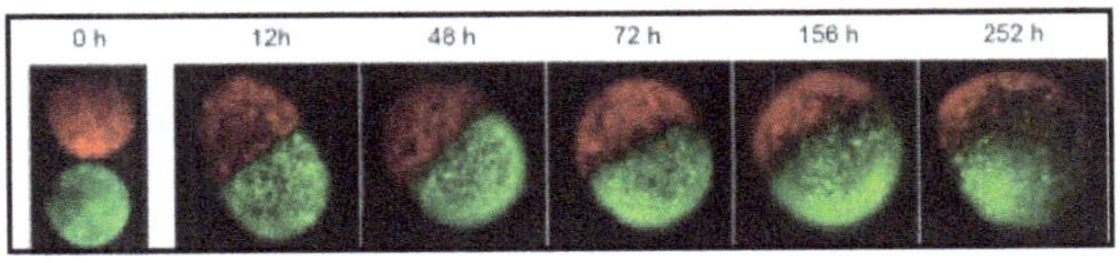

Shown in the picture above is an invasion assay of magnetically levitated multicellular spheroids. Fluorescence images of human glioblastoma (GBM) cells (green; GFP-expressing cells) and normal human astrocytes (NHA) (red; mCherry-labelled) cultured separately and then magnetically guided together (left, time 0). Invasion of GBM into NHAin 3D culture provides a powerful new assay for basic cancer biology and drug screening (right, 12h to 252h).

Vascular Simulation with Stem Cells

By facilitating assembly of different populations of cells using the MLM, consistent generation of organoids termed adipospheres capable of simulating the complex intercellular interactions of endogenous white adipose tissue (WAT) can be achieved.

Co-culturing 3T3-L1 pre-adipocytes in 3D with murine endothelial bEND.3 cells creates a vascular-like network assembly with concomitant lipogenesis in perivascular cells.

In addition to cell lines, WAT organogenesis can be simulated from primary cells.

Adipocyte-depleted stromal vascular fraction (SVF) containing adipose stromal cells (ASC), endothelial cells, and infiltrating leukocyte derived from mouse white adipose tissue (WAT) were cultured in 3D. This revealed organoids striking in hierarchical organization with distinct capsule and internal large vessel-like structures lined with endothelial cells, as well as perivascular localization of ASC.

Upon adipogenesis induction of either 3T3-L1 adipospheres or adipospheres derived from SVF, the cells efficiently formed large lipid droplets typical of white adipocytes in vivo, whereas only smaller lipid droplet formation is achievable in 2D. This indicates intercellular signaling that better recapitulates WAT organogenesis.

This MLM for 3D co-culturing creates adipospheres appropriate for WAT modeling ex vivo and provides a new platform for functional screens to identify molecules bioactive toward individual adipose cell populations. It can also be adopted for WAT transplantation applications and aid other approaches to WAT-based cell therapy.

Organized Co-culturing to Create in Vivo-like Tissue

Using the MagPen™ (a Nano3D Biosciences, Inc. product), organized 3D co-cultures similar to native tissue architecture can be rapidly created. Endothelial cells (PEC), smooth muscle cells (SMC), fibroblasts (PF), and epithelial cells (EpiC) cultured with the Bio-Assembler™ can be sequentially layered in a drag-and-drop manner to create bronchioles that maintain phenotype and induce extracellular matrix formation.

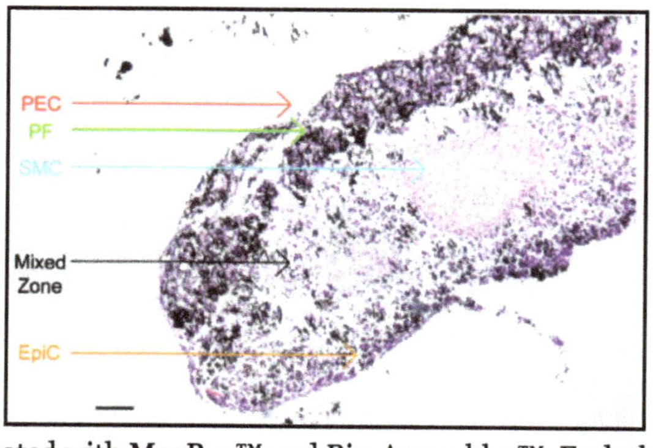

Organized bronchiole created with MagPen™ and Bio-Assembler™. Endothelial cells (PEC), smooth muscle cells (SMC), fibroblasts (PF), and epithelial cells (EpiC) can be sequentially layered. Scale bar: 100um.

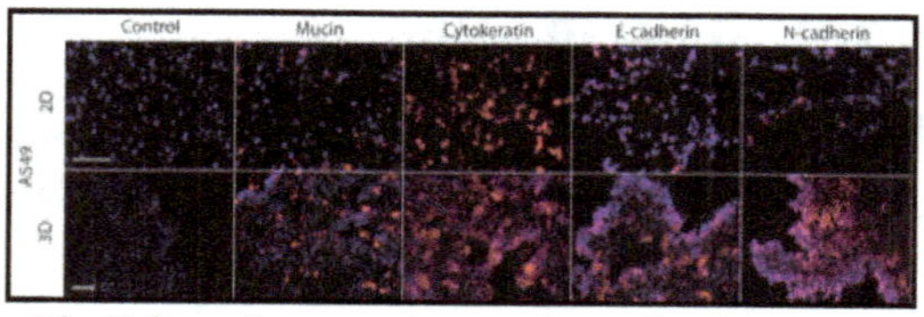

IHC reveals epithelial markers on A549 after culturing in 3D with the MLM.

Cell Types Cultured

Listed below are the cell types (primary and cell lines) that have been successfully cultured by the magnetic levitation method. The second table is the same but with images included. More images are available at Nano3D Biosciences, Inc.

Cells	Cell line	Organism	Organ tissue
Murine endothelial	Cell line	Mouse	Vessel
Murine adipocyte	Cell line	Mouse	Adipose
Rattus norvegicus hepatoma	Cell line	Rat	Liver
Pulmonary fibroblasts (HPF)	Primary	Human	Lung
Pulmonary endothelial (HPMEC)	Primary	Human	Lung
Small airway epithelial (HSAEpiC)	Primary	Human	Lung
Bronchial epithelial	Primary	Human	Lung
Human alveolar adenocarcinoma	A549	Human	Lung
Type II alveolar	Primary	Human	Lung
Tracheal smooth muscle (HTSMC)	Primary	Human	Lung
Mesenchymal stem cells (HMSC)	Primary	Human	Bone marrow
Bone marrow endothelial cells (HBMEC)	Primary	Human	Bone marrow
Dental pulp stem cells	Primary	Human	
Human umbilical vein endothelial cells (HUVEC)	Primary	Human	
Murine chondrocytes	Primary	Mouse	Bone
Murine adipose tissue	Primary	Mouse	
Heart valve endothelial	Primary	Porcine	
Pre-adipocytes fibroblasts	3T3	Mouse	
Neural stem cells	C17.2	Mouse	Brain
Human embryonic kidney cells	HEK293	Human	Kidney
Melanoma	B16	Mouse	Skin
Astrocytes	NHA	Human	Brain
Glioblastomas	LN229	Human	Brain
T-cells and antigen presenting cells		Human	
Mammary epithelial	MCF10A	Human	Breast
Breast cancer	MDA231	Human	Breast
Osteosarcoma	MG63	Human	Bone

Protoplast Isolation and Regeneration

The term protoplast was introduced by Hanstein in 1880. It refers to the cellular content excluding cell wall or can also be called as naked plant cell. It is described as living matter enclosed by a plant cell membrane. Protoplast isolation for the first time was carried out by Klercker in 1892 using mechanical method on the plasmolysed cells. The application of protoplast technology for the improvement of plants offers fascinating option to complement conventional breeding programs. The ability of isolated protoplasts to undergo fusion and take up macromolecules and cell organelles offers many possibilities in genetic engineering and crop improvement (Bhojwani et al . 1977). The experiments involving protoplasts consist of three stages:

i. protoplast isolation.

ii. protoplast fusion (leading to gene uptake).

iii. development of regenerated fertile plants from the fusion product (Hybrid).

Depending upon the species and culture conditions, the protoplasts may have the potential to:

i. regenerate a cell wall.

ii. dedifferentiate to form callus.

iii. divide mitotically and proliferate clonally.

iv redifferentiate into shoots, roots or embryos and produce a complete plantlet.

However, to fully explore the potentials for protoplast-technology, efficient and reproducible methods for protoplast isolation and purification must first be established. Since leaf tissue is a readily accessible source of genetically uniform cells, it is often desirable to use mesophyll protoplasts in somatic hybridization studies, but, leaf tissues, in general, do not yield large number of protoplasts owing to the difficulty in removing the lower epidermis (Chaturvedi 2003). An alternative, therefore, is the cultured cell material where protoplasts can show greater potential to divide (Bhojwani and Razdan 1996).

Protoplast Isolation

Protoplast isolation may be carried out by Mechanical disruption method or enzymatic method. Out of these two methods, enzymatic method is preferred as it provides better protoplast yield with low tissue damage while mechanical method causes maximum tissue chopping with lower protoplast yields. Both of these methods are described below:

Mechanical method:

Klercker in 1892 pioneered the isolation of protoplasts by mechanical methods. In this method, the cells were kept in suitable plasmolyticum, for example CPW containing 13% w/v mannitol. Once the plasmolysis is complete, while remaining in the osmoticum, the leaf lamina would be cut with a sharp-edged knife. In this process some of the plasmolyzed cells were cut only through the cell wall, releasing intact protoplasts while some of the protoplasts may be damaged inside many cells. Protoplasts that were trapped in a cell and only the corner had been cut off could be encouraged to come out by reducing the osmolarity slightly to force the protoplasts swell to force their way out of the cut surface. The released protoplasts then have to be separated from damaged ones and cell debris.

Disadvantages

- Lower protoplast yield.
- Labour intensive method.
- Protoplast obtained has low viability.
- Method is applicable only to vacuolated cells.

Enzymatic method:

In 1960, E.C. Cocking demonstrated the possibility of enzymatic isolation of a large number of protoplasts from cells of higher plants. This method involves leaf sterilization followed by peeling of the lower epidermis to release cells which are plasmolyzed and added to enzyme mixture followed by harvest of protoplast as shown in. Either of the procedures for enzymatic isolation can be used: sequential enzymatic hydrolysis or mixed enzymatic hydrolysis.

In the sequential isolation, firstly, cells are separated by the use of a maceration enzyme – a pectin hydrolyzing enzyme such as, macerozyme or Pectolyase. Once the cells are separated, they are washed in CPW solution free of enzymes but containing plasmolyticum by gentle centrifugation (100g). The pellet is retained and resuspended in the second enzyme like, cellulases and hemicellulases, used to hydrolyse the remaining cell was component. Once the protoplasts are released they are washed with CPW to remove the debris.

In the mixed enzymatic approach, Plant tissues are plasmolyzed in the presence of a mixture of pectinases and cellulases, thus, inducing simultaneous separation of cells and degradation of their walls to release the protoplasts directly in a single step.

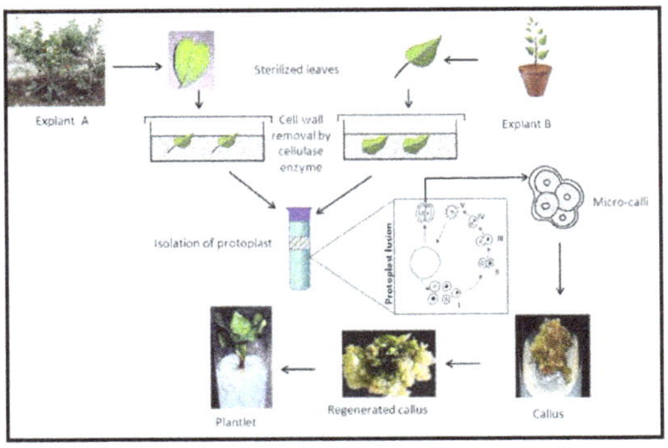

Steps involved in protoplast isolation, fusion and regeneration.

Conditions required for enzyme activity :

- Enzymes are pH and temperature dependent, thus, for enzymatic release of protoplast an enzyme showing activity at pH range 4.7-6.0 and temperature range of 25-30°C is used.
- Duration of enzyme pretreatment and condition of light presence required for incubation may also be determined.
- Enzyme mixture used should essentially consist of cellulose, hemicellulase and pectinase which facilitate the degradation of cellulose, hemicelluloses and pectin, respectively.
- The concentration of sugar alcohols used as osmoticum (mannitol) must be empirically defined.

Factors Affecting Yield and Viability of Protoplasts

i. *Source of material* : Leaves were the most convenient source of the plant protoplasts because it allows the isolation of a large number of relatively uniform cells without killing the plants. Moreover the mesophyll cells are loosely arranged, the enzymes have an easy access to the cell wall. The parent plant age and the conditions in which it is growing have profound effect on the yield of protoplast. Due to the difficulty in isolating culturable protoplast from leaf cells of cereals and some other species their cultured cells can be used as a source material. The yield of protoplasts depends upon the growth rate and growth phase of the cells. Generally embryogenic suspension cultures are used to obtain totipotent protoplasts.

ii. *Pre-enzyme treatments* : To facilitate the penetration of enzyme solution into the intercellular spaces of leaf, which is essential for effective digestion, various methods are followed. The most commonly used method is to peel the lower epidermis and float the stripped pieces of leaf on the enzyme solution in a manner that the peeled surface is in a contact with the solution. Most of the time it is not convenient to peel the epidermis, in such cases cutting the leaf or tissue into small strips (1- 2 mm wide) has been found useful. When combined with vacuum infiltration the latter approach has proved very effective. Brushing of leaves with a soft brush or with the cutting edge of a scalpel may also improve the enzymatic action. Large calli are chopped into pieces and can be transferred to enzyme mixture. Agitation of incubation mixture during enzyme treatment improves protoplast yield from cultured cells.

iii. *Enzyme treatment* : The release of protoplast is very much dependent on the nature and concentration of the enzymes used. The two major enzymes required for the isolation of protoplast are cellulase and pectinase. The cellulase is required to digest the cellulosic cell walls and the pectinase mainly degrades the middle lamella. Some of the tissues may require other enzymes like, hemicellulase, driselase, macerozyme and pectolyase. The activity of enzyme is pH dependent. The pH of the enzyme solution is adjusted somewhere between 4.7 to 6.0.

Choice of Plasmolyticum

The two most commonly used compounds are the sugar alcohols - mannitol and sorbitol. Of these, mannitol is the most preferred since it is not metabolized by the plant cells. Once the protoplast divides and regenerates the cell wall, no more osmoticum is required. It is, therefore, should be removed gradually from the medium otherwise cell division stops. To slowly remove the osmoticum from the medium, the protoplast can be isolated in a high osmoticum mixture consisting of both mannitol and sucrose, the sucrose will be metabolized by the dividing protoplasts and thus, will reduce the osmolarity of the medium. Normally, mannitol is used at concentration range of 11-13%.

A solution into which the osmoticum is often, but not always, added is called CPW salts mix or CPW for short. This has been observed much more beneficial than using distilled water as a solvent in obtaining high yields of viable protoplasts:

Table: Salt mix of protoplast washing media solution (Cocking, Peberdy and White – CPW).

KH_2PO_4	27.2 mg
KNO_3	101 mg
$CaCl_2.2H_2O$	1480 mg
$MgSO_4.7H_2O$	246 mg
KI	0.16 mg
$CuSO_4.5H_2O$	0.025 mg
Made to 1 litre with water : pH 5.8	

Although CPW is most widely used solution into which osmoticum or enzymes are added, some times culture medium used to grow cells or plants can also be utilized for protoplast isolation at one tenth concentration. Low concentration of culture medium is much more advantageous when compared with CPW.

Protoplast Purification

Enzyme treatment results in suspension of protoplast, undigested tissues and cellular debris. This suspension is passed through a metal sieve or a nylon mesh (50-100 μm) in order to remove undigested cellular clumps. The filtered protoplast-enzyme solution is mixed with a suitable volume of osmoticum, solution is centrifuged to pellet the protoplasts, pellet of protoplast is resuspended in osmoticum of similar concentration as used in enzyme mixture. The protoplast band is sucked in Pasteur pipette and is put into other centrifuge and finally suspended in culture medium at particular density; this is explained by the figure.

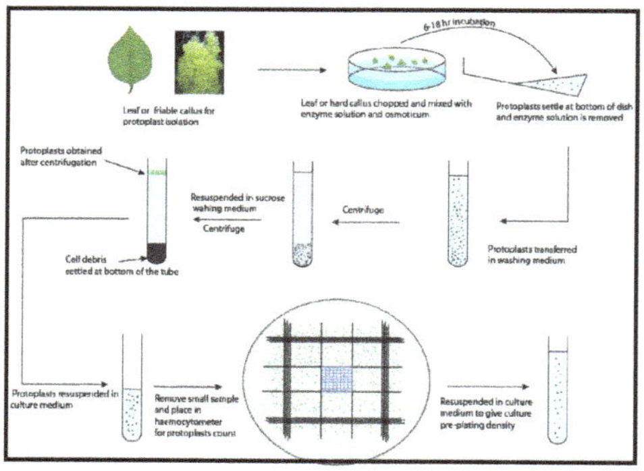

schematic diagram of protoplast purification.

Protoplast viability: The isolated protoplast must have a spherical shape when observed by a light microscope, protoplast can be stained using following stain:

- *Fluorescein diacetate staining method:* FDA accumulates inside the plasma-lemma of viable protoplasts. Live protoplasts contain esterases which cleave FDA to release fluorescein which fluoresces yellowish-green using fluorescence microscopy within 5 min. FDA dissociates from membrane after about 15 min. It is used at a concentration of 0.01% dissolved in acetone.
- *Calcofluor White staining:* This staining method assures protoplast viability by detecting onset of cell wall formation. Calcofluor binds to beta linked glucosides in newly synthesized cell wall which can be observed as a fluorescent ring around the membrane. Optimum staining is achieved when 0.1 ml of protoplast is mixed with 5.0 µl of 0.1% w/v solution of CFW.
- *Protoplast viability* can also be detected by monitoring oxygen uptake of cells by oxygen electrode, which shows respiration.
- *Variation of protoplast size* with changing osmotic concentration also enables viability of protoplast.

Protoplast Culture

Protoplasts Culture Techniques

The culture requirements and the culture methods are same for both protoplasts and single cells. The main difference is the requirement of suitable osmoticum for protoplasts until they regenerate a strong wall. Isolated protoplasts are either cultured in liquid or semisolid agar or agarose media plates, sometimes the protoplast is first grown in liquid media and then transferred into the agar media plates. The following techniques have been adopted in order to maintain number of protoplast population between minimum and maximum effective densities after plating up:

i. *Liquid method* : This method is preferred in earlier stages of culture as it provides (a) easier dilution and transfer, (b) the osmotic pressure of liquid media can be effectively reduced after a few days of culture (c) the cell density can be reduced or cells of special interest can be isolated easily. In Liquid medium, the protoplast suspension is plated as a thin layer in petriplates, incubated as static culture in flasks or distributed in 50-100 μl drops in petriplates and stored in a humidifier chamber.

ii. *Embedded in Agar/ Agarose* : Agarose is a preferred choice in place of agar and this has improved the culture response. This method of agar culture keeps protoplast in fixed position, thus, prevents it from forming clumps. Immobilized protoplasts give rise to clones which can then be transferred to other media. In practice, the protoplasts suspended in molten (40°C) agarose medium (1.2% w/v agarose) are dispensed (4ml) into small (3.5-5cm diameter) plates and allowed to solidify. The agarose layer is then cut into 4 equal sized blocks and transferred to larger dishes (9 cm diameter) containing liquid medium of otherwise the same composition. Alternatively, protoplasts in molten agarose medium are dispensed as droplets (50-100 μl) on the bottom of petri plates and after solidification the droplets are submerged in the same liquid medium.

iii. *Feeder layer* : In order to culture protoplast at low density, a feeder layer technique is adopted. A feeder layer of X-ray irradiated non-dividing but metabolically active protoplasts after washing are plated in soft agar medium. Non-irradiated protoplasts of low density are plated over this feeder layer. The protoplasts of the same species or different species can be used as a feeder layer.

iv. *Co-culturing* : This method involves co-culture of protoplasts from two different species to promote their growth or that of the hybrid cells. Metabolically active and dividing protoplasts of two types - slow and fast growing are cultured together, the fast growing protoplast provide other species with diffusible chemicals and growth factors which helps in cell wall formation and cell division. The co-culture methods is generally used where calli arising from two types of protoplasts can be morphologically distinguished. For example, protoplasts isolated from albino plants and green plants are easily distinguishable based on color where albino protoplast will develop non green colonies.

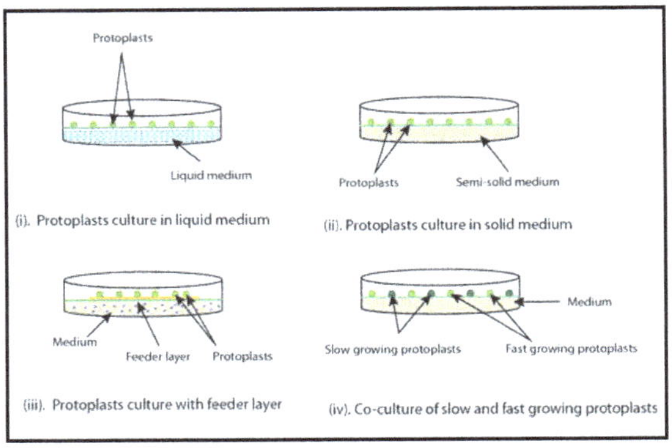

protoplast culture techniques.

Culture Medium

The nutritional requirement of protoplast is almost similar to that of the cultured plant cells. Mostly the salts of MS (Murashige and Skoog, 1962) and B_5 (Gamborg et al. 1968) media and their modifications have been used. Ammonium salts have been found detrimental to protoplasts survival of many species, and media have been devised that either have a reduced concentration of ammonium or lack it. Concentration of zinc is reduced while the concentration of calcium is increased as it enhances the membrane stability. Osmolarity is maintained by addition of sorbitol, mannitol, glucose or sucrose and mannitol being widely used osmoticum as it is not used by the dividing cells, thus, maintains the osmolarity of the medium.

Glucose is preferred carbon source as sucrose do not satisfy protoplast culture. One or two amino acids are added at low concentration. Growth regulators are required essentially in protoplast culture generally high concentration of auxins (NAA, 2,4-D) along with lower concentration of cytokinins (BAP, Zeatin) is used.

Environmental conditions: High light intensity inhibits growth of protoplast hence initially protoplast is grown in dim light for few days and then transferred to light of about 2000-5000 lux. However, better results are obtained when cultured in darkness.

Plating Density

Like cell cultures, the initial plating density of protoplasts has profound effect on plating efficiency. Protoplasts are cultured at a density of 1×10^4 to 1×10^5 protoplasts ml^{-1} of the medium. At high density the cell colonies arising from individual protoplasts tend to grow into each other resulting into chimera tissue if the the protoplast population is genetically heterogeneous. Cloning of individuals cells, which is desirable in somatic hybridization and mutagenic studies, can be achieved if protoplasts or cells derived from them can be cultured at a low density.

Protoplast Development and Regeneration

Protoplast starts to regenerate a cell wall within few days (2-4 days) of culture and during this process, protoplasts lose their characteristic spherical shape which has been taken as an indication of new wall regeneration. Cell wall regeneration can be confirmed by Calcofluor White staining method. There is direct relationship between wall formation and cell division. Protoplasts which are not able to regenerate a proper wall fail to undergo normal mitosis. Protoplasts with a poorly developed wall often show budding and may enlarge several times their original volume. They may become multinucleate because karyokinesis is not accompanied by cytokinesis. Among other reasons, inadequate washing of the protoplasts prior to culture leads to these abnormalities.

And completes process when provided with suitable condition of light, pH and temperature newly synthesized protoplast can be visualized by staining. Once the cell wall formation is completed, cells undergo division resulting in increase size of cells. After an interval of 3 weeks, small cell colonies appear, these colonies are transferred to an osmotic-free callus induction medium. This is followed by introduction into organogenic or embryogenic medium leading to plantlet development.

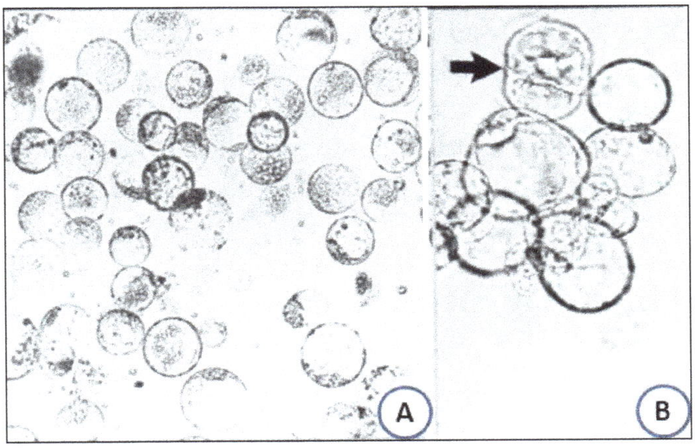

Protoplast isolation and cell wall regeneration. A. Isolated protoplast showing spherical structure; B. Wall is regenerated around the protoplast and one of the protoplasts showing cell division (arrow marked).

Somatic Hybridization and Cybridization

Sexual hybridization since time immemorial has been used as a method for crop improvement but it has its own limitations as it can only be used within members of same species or closely related wild species. Thus, this limits the use of sexual hybridization as a means of producing better varieties. Development of viable cell hybrids by somatic hybridization, therefore, has been considered as an alternative approach for the production of superior hybrids overcoming the species barrier. The technique can facilitate breeding and gene transfer, bypassing problems associated with conventional sexual crossing such as, interspecific, intergeneric incompatibility. This technique of hybrid production via protoplast fusion allows combining somatic cells (whole or partial) from different cultivars, species or genera resulting in novel genetic combinations including symmetric somatic hybrids, asymmetric somatic hybrids or somatic cybrids.

The most common target using somatic hybridization is the gene of symmetric hybrids that contain the complete nuclear genomes along with cytoplasmic organelles of both parents. This is unlike sexual reproduction in which organelle genomes are generally contributed by the maternal parent. On the other hand, somatic cybridization is the process of combining the nuclear genome of one parent with the mitochondrial and/

or chloroplast genome of a second parent. Cybrids can be produced by donor-recipient method or by cytoplast-protoplast fusion. Incomplete asymmetric somatic hybridization also provides opportunities for transfer of fragments of the nuclear genome, including one or more intact chromosomes from one parent (donor) into the intact genome of a second parent (recipient).

These methods provide genetic manipulation of plants overcoming hurdle of sexual incompatibility, thereby, serving as a method of bringing together beneficial traits from taxonomically distinct species which cannot be achieved by sexual crosses. Several parameters such as, source tissue, culture medium and environmental factors influence ability of a protoplast derived hybrid cells to develop into a fertile plant. The general steps involved in somatic hybridization and cybridization methods are elaborated in Figur.

Steps Involved in Somatic Hybridization

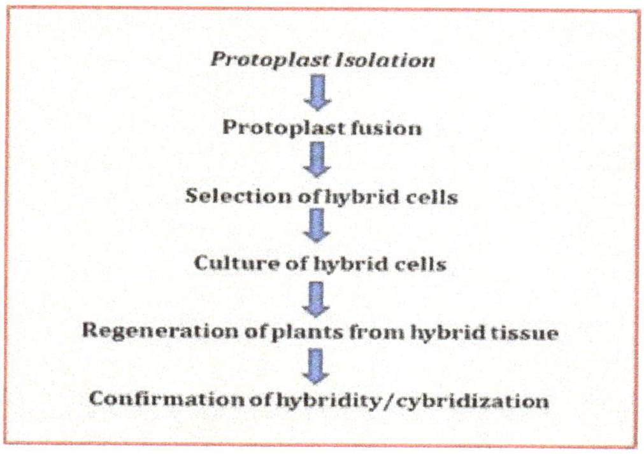

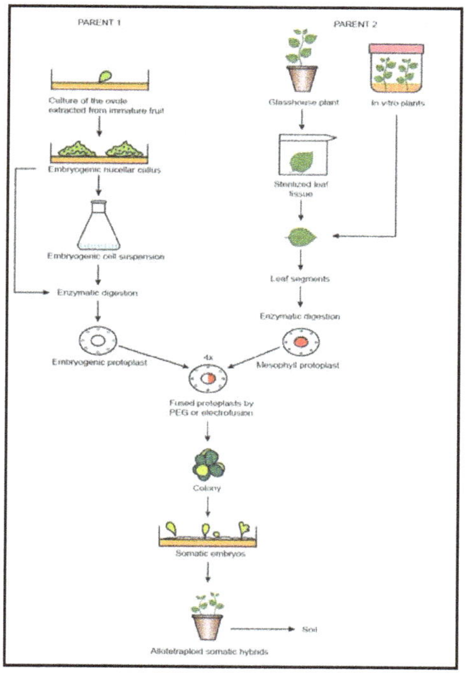

Schematic view of symmetric protoplast fusion producing somatic hybrids.

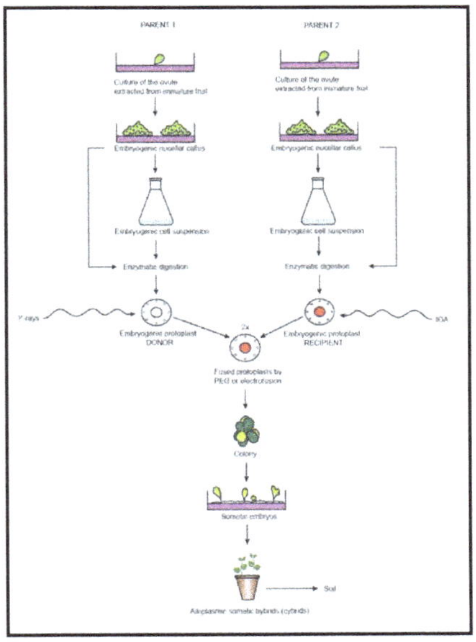

Schematic view of asymmetric protoplast fusion using donor-recipient method resulting into creation of alloplasmic somatic hybrid or cybrids.

Protoplast Fusion

Protoplast fusion could be spontaneous during isolation of protoplast or it can be induced by mechanical, chemical and physical means. During spontaneous process, the adjacent protoplasts fuse together as a result of enzymatic degradation of cell walls forming homokaryons or homokaryocytes, each with two to several nuclei. The occurrence of multinucleate fusion bodies is more frequent when the protoplasts are prepared from actively dividing callus cells or suspension cultures. Since the somatic hybridization or cybridization require fusion of protoplasts of different origin, the spontaneous fusion has no value. To achieve induced fusion, a suitable chemical agent (fusogen) like, $NaNO_3$, high Ca^{2+}, polyethylene glycol (PEG), or electric stimulus is needed.

i. Fusion by means of $NaNO_3$: It was first demonstrated by Kuster in 1909 that the hypotonic solution of $NaNO_3$ induces fusion of isolated protoplast forming heterokaryon (hybrid). This method was fully described by Evans and Cocking (1975), however this method has a limitation of generating few no of hybrids, especially when highly vacuolated mesophyll protoplasts are involved.

ii High pH and Ca^{++} treatment: This technique lead to the development of intra- and interspecific hybrids. It was demonstrated by Keller and Melcher in 1973. The isolated protoplasts from two plant species are incubated in 0.4 M mannitol solution containing high Ca^{++}(50 mM $CaCl_2.2H_2O$) with highly alkaline pH of 10.5 at 37°C for about 30 min. Aggregation of protoplasts takes place at once and fusion occurs within 10 min.

iii Polyethylene glycol treatment: Polyethylene glycol (PEG) is the most popularly known fusogen due to ability of forming high frequency, binucleate heterokaryons with low cytotoxicity. With PEG the aggregation occurred mostly between two to three protoplasts unlike Ca^{++} induced fusion which involves large clump formation. The freshly isolated protoplasts from two selected parents are mixed in appropriate proportions and treated with 15-45% PEG (1500-6000MW) solution for 15-30 min followed by gradual washing of the protoplasts to remove PEG. Protoplast fusion occurs during washing. The washing medium may be alkaline (pH 9-10) and contain a high Ca^{++} ion concentration (50 mM). This combined approach of PEG and Ca^{++} is much more efficient than the either of the treatment alone. PEG is negatively charged and may bind to cation like Ca^{++}, which in turn, may bind to the negatively charged molecules present in plasma lemma, they can also bind to cationic molecules of plasma membrane. During the washing process, PEG molecules may pull out the plasma lemma components bound to them. This would disturb plamalemma organization and may lead to the fusion of protoplasts located close to each other. The technique is nonselective thus, induce fusion between any two or more protoplasts.

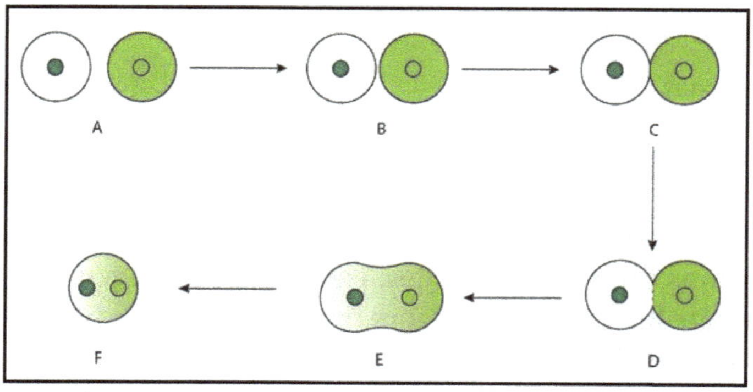

Sequential stages in protoplast fusion. (A) two separate protoplasts, (B) agglutination of two protoplasts, (C and D) Membrane fusion at localized site, and (E and F) development of spherical heterokaryon.

iv. Electrofusion: The chemical fusion of plant protoplast has many disadvantages – (1) The fusogen are toxic to some cell systems, (2) it produces random, multiple cell aggregates, and (3) must be removed before culture. Compare to this, electrofusion is rapid, simple, synchronous and more easily controlled. Moreover, the somatic hybrids produced by this method show much higher fertility than those produced by PEG-induced fusion.

Zimmermann and Scheurich (1981) demonstrated that batches of protoplasts could be fused by electric fields by devising a protocol which is now widely used. This protocol involves a two-step process. First, the protoplasts are introduced into a small fusion chamber containing parallel wires or plates which serve as electrodes. Second, a low-voltage and rapidly oscillating AC field is applied, which causes protoplasts to become aligned into chains of cells between electrodes. This creates complete cell-to-cell contact within a few minutes. Once alignment is complete, the fusion is induced by application of a brief spell of high-voltage DC pulses (0.125-1 $kVcm^{-1}$). A high voltage DC pulses induces a reversible breakdown of the plasma membrane at the site of cell contact, leading to fusion and consequent membrane reorganization. The entire process can be completed within 15 min.

Selection of Fusion Products

The somatic hybridization by electrofusion of protoplasts allow one-to-one fusion of desired pairs of protoplasts and, therefore, it is easy to know the fate of fusion products. However, protoplast suspension recovered after chemical treatments (fusogen) consists of the following cell types:

i. unfused protoplasts of the two species/strains.

ii. products of fusion between two or more protoplasts of the same species (homokaryons).

iii. hybrid protoplasts produced by fusion between one (or more) protoplasts of each of the two species (heterokaryons).

The heterokaryons which are the potential source of future hybrids constitute of a very small (0.5-10%) proportion of the mixture. Therefore, an effective strategy has to be employed for their identification and isolation. Various protocols have been proposed and practiced for the effective selection of hybrids, including morphological basis, complementation of biochemical and genetic traits of the fusing partners, and manual or electronic sorting of heterokaryons/hybrid cells.

Morpho-physiological basis: The whole mixture of the protoplasts are cultured after fusion treatment and the resulting calli or regenerants are screened for their hybrid characteristics. Occasionally the hybrid calli outgrow the parental cell colonies and are identified by their intermediate morphology, i.e. green with purple coloured cells. However, the process is labour intensive and requires glasshouse facilities. It is limited to certain combinations showing differences in their regeneration potential under specific culture conditions.

Complementation: In this case complementation or genetic or metabolic deficiencies of the two fusion partners are utilized to select the hybrid component. When protoplasts of two parents, (one parent bearing cytoplasmic albino trait and the other parent bearing green trait) each parent carrying a non-allelic genetic or metabolic defect are fused, it reconstitutes a viable hybrid cell, of wild type in which both defects are mutually abolished by complementation, and the hybrid cells are able to grow on minimal medium non-permissive to the growth of the parental cells bearing green trait. Later, the calli of hybrid nature could be easily distinguished from the parental type tissue (albino trait) by their green color. The complementation selection can also be applied to dominant characters, such as dominant resistance to antibiotics, herbicides or amino acid analogues.

Isolation of heterokaryons or hybrid cells: The manual or electronic isolation of heterokaryons or hybrid cells is the most reliable method. Manual isolation requires that the two parental type protoplasts have distinct morphological markers and are easily distinguishable. For example, green vacuolated, mesophyll protoplasts from one parent and richly cytoplasmic, non green protoplasts from cultured cells of another parent. The dual fluorescence method also helps easy identification of fusion products. In this case, the protoplast labeled green by treatment with fluorescein diacetate (FDA, 1-20 mgl^{-1}) are fused with protoplasts emitting a red fluorescence, either from chlorophyll autofluorescence or from exogenously applied rhodamine isothiocyanate (10-20 mgl^{-1}). The labeling can be achieved by adding the compound into the enzyme mixture. This can be applied even for morphologically indistinguishable protoplasts from two parents.

Verification and Characterization of Somatic Hybrids

As no system is foolproof and they have their own advantages and disadvantages. Therefore, even after selecting the desired hybrids/cybrids following protoplast fusion, it is required to carry out one or more tests to compare the parent protoplast lines with the putative hybrids. Some of the techniques that can be tried are:

Morphology: Somatic hybrids in most of the cases show characters intermediate between the two parents such as, shape of leaves, pigmentation of corolla, plant height, root morphology and other vegetative and floral characters. The method is not much accurate as tissue culture conditions may also alter some morphological characters or the hybrid may show entirely new traits not shown by any of the parents.

Isozyme analysis: Multiple molecular forms of same enzyme which catalyses similar or identical reactions are known as isozymes. Electrophoresis is performed to study banding pattern as a check for hybridity. If the two parents exhibit different band patterns for a specific isozyme the putative hybrid can be easily verified. The isozymes commonly used for hybrid identification include, acid phosphatase, esterase, peroxidase.

Cytological analysis: Chromosome counting of the hybrid is an easier and reliable method to ensure hybridity as it also provides the information of ploidy level. Cytologically the chromosome count of the hybrid should be sum of number of chromosomes from both the parents. Besides number of chromosomes, the size and structure of chromosomes can also be monitored. However, the approach is not applicable to all species, particularly where fusion involves closely related species or where the chromosomes are very small. Moreover, sometimes the somaclonal variations may also give rise to different chromosome number.

Molecular analysis: Specific restriction pattern of nuclear, mitochondrial and chloroplast DNA characterizes the plastomes of hybrids and cybrids. Molecular markers such as RFLP, RAPD, ISSR can be employed to detect variation and similarity in banding pattern of fused protoplasts to verify hybrid and cybrid.

Cybrids or Cytoplasmic Hybrids:

Sexual hybridization involves fusion of the nuclear genes of both the parents but somatic hybrids involves even cytoplasm from both the parental species in hybrid obtained by protoplast fusion. However, in another case somatic hybrids containing nuclear genome of one parent but cytoplasm from both the parents, are termed as cybrids. The approach is time consuming and require several years of crossing plants provides an opportunity to study interparental mitochondrial, chloroplast fusion giving rise to plants with novel genomes.

Methods to produce cybrids: They are produced in variable frequencies in normal protoplast fusion experiments due to one of the following methods:

1. Fusion of normal protoplast with an enucleated protoplast. The enucleated protoplast can be produced by high speed centrifugation (20,000-40,000xg) for 60 min with 5-50% percoll.
2. Fusion between a normal protoplast and another protoplast with a non-viable nucleus or suppressed nucleus.
3. Elimination of one of the nuclei after heterokaryons formation.
4. Selective elimination of chromosomes at a later stage.
5. Irradiating (with X-rays or gamma rays) the protoplasts of one species prior to fusion in order to inactivate their nuclei.
6. By preparing enucleate protoplasts (cytoplasts) of one species and fusing them with normal protoplasts of the other species.

Cybrids provide the following unique opportunities: (i) transfer of plasmogenes of one species into the nuclear background of another species in a single generation, and even in (ii) sexually incompatible combinations, (iii) recovery of recombinants between the parental mitochondrial or chloroplast DNAs (genomes), and (iv) production of a wide variety of combinations of the parental and recombinant chloroplasts with the parental or recombinant mitochondria.

Applications of Somatic Hybridization

1. Novel interspecific and intergeneric crosses which are difficult to produce by conventional methods can be easily obtained.
2. Important characters, such as resistance to diseases, ability to undergo abiotic stress and other quality characters, can be obtained in hybrid plant by the fusion of protoplasts of plant bearing particular character to the other plant which may be susceptible to diseases.

3. Protoplasts of sexually sterile haploid, triploid, aneuploid plants can be fused to obtain fertile diploids and polyploids.
4. Studying cytoplasmic genes may be helpful to carry out plant breeding.
5. Most of the agronomically important traits, such as cytoplasmic male sterility, antibiotic resistance and herbicide resistance, are cytoplasmically encoded, hence can be easily transferred to other plant.
6. Plants in juvenile stage can also be hybridized by means of somatic hybridization.
7. Somatic hybridization can be used as a method for the production of autotetraploids.

Limitations of Somatic Hybridization

1. Application of protoplast methodology requires efficient plant regeneration system from isolated protoplasts. Protoplasts from two species can be fused, however, production of somatic hybrids is not easy.
2. Lack of a proper selection method for fused products (hybrids) poses a problem.
3. The end product of somatic hybridization are often unbalanced (sterile, misformed and unstable).
4. Somatic hybridization of two diploids leads to formation of amphidiploids which is unfavorable.
5. It is not sure for a character to completely express after somatic hybridization.
6. The regeneration products of somatic hybridization are often variable due to somaclonal variation, chromosome elimination, organelle segregation.
7. All diverse intergeneric somatic hybrids are sterile and, therefore, have limited chances of development of new varieties.
8. To transfer useful genes from wild species to cultivated crop, it is necessary to achieve intergeneric recombination or chromosome substitution between parental genomes.

Cryopreservation

In recent years, with the enormous increase in the population, pressure on the forest and the land resources have increased. This results in depletion of population of medicinal and aromatic plant species. Even some of the plant species are at the verge of vanishing from the forest. The list of endangered species is growing day by day. The conventional methods of germplasm preservation are prone to possible catastrophic losses because of:

i. Attack by pathogen and pests.

ii. Climatic disorders.

iii. Natural disasters.

iv. Political and economic causes.

In addition, the seeds of many important medicinal plants lose their viability in a short time under conventional storage system.

The conservation of germplasm can be done by two methods:

1. In-situ preservation: Preservation of the germplasm in their natural environment by establishing biospheres, national parks etc.

2. Ex-situ preservation: In the form of seeds or by *in vitro* cultures.

Seeds form the most common material to conserve plant germplasm, however, the method has the following disadvantages:

i. Discrete clones cannot be maintained in the form of seeds.

ii. Some plants do not produce fertile seeds.

iii. Loss of seed viability.

iv. Seed destruction by pests, etc.

v. Poor germination rate.

vi. This is useful for seed propagating plants and is not applicable to vegetatively propagated crops, like potato, ginger etc.

In vitro preservation by tissue culture has several advantages over seed preservation:

i. Large amount of materials can be stored in a small area.

ii. The material could serve as an excellent form of nucleus stock to propagate large number of plants rapidly, when required.

iii. Under special storage conditions the plants do not require frequent splitting and pruning.

iv. Being free from known viruses and pathogens, the clonal plant material could be sent from country to country, thus, minimizing the obstructions imposed by quarantine systems on the movement of live plants across national boundaries.

v. Protection from natural hazards.

vi. The plants are not exposed to the threat of changing government policies and urban development.

There are few disadvantages of in vitro system to be used for conservation of plant material:

i. It is a costly process.

ii. In cultures, plants can be maintained by serial subcultures at frequent intervals for virtually unlimited periods. However, the storage of germplasm by serial subcultures risks the loss of plant material by microbial contamination due to human error and also, is uneconomical. Moreover, in long-term callus and suspension cultures, the regeneration potential, biosynthetic properties and genetic make-up of the cells suffer. The basic requirement of a plant tissue culture method is the preservation of genetic resources, therefore, is to reduce the frequency of subcultures to a bare minimum.

Cryopreservation means preservation in the frozen state. The principle involved in cryopreservation is to bring the plant cell and tissue cultures to zero metabolism or non-dividing state by mean of storage of germplasm at a very low temperatures, (i) Over solid CO_2 (-79°C), (ii) Deep freezers (-80°C), (iii) in vapor phase nitrogen (-150°C), (iv) in liquid nitrogen (-196°C).

Among these, the most commonly used non-lethal storage of biological material at ultra-low temperature is by employing liquid nitrogen. At the temperature of liquid nitrogen (-196°C), almost all the metabolic activities of cells are ceased and the sample can then be preserved in such state for extended periods.

Steps Involved in Cryopreservation

The technique of cryopreservation involves the following steps:

i. Selection of plant material.

ii. Pre-culture.

iii. Cryoprotective treatment.

iv. Freezing and storage.

v. Thawing.

vi. Reculture.

Selection of Plant Material

The morphological and physiological conditions of the plant material, prior to freezing, considerably influence its ability to survive freezing at -196°C. Generally, small, richly cytoplasmic and meristematic cells survive better than the larger, highly vacuolated cells. Therefore, cell suspensions should be frequently subcultured and frozen in the late lag phase or exponential phase when the majority of the cells are in the preferred condition. While preservation of cell lines remains useful with respect to *in vitro* production of secondary metabolites, cultured cells are not the ideal system for germplasm storage. Instead, organized structures, such as shoot apices, embryos or young plantlets are preferred. The reasons to shift from cell cultures to organized cultures are as follows:

i. The genetic instability of cells in long term callus and cell suspension cultures is a very common phenomenon and there is no effective measure to control it so far. Moreover, most of the callus cultures are initiated from non-meristematic cells of the plant body which might exhibit polysomaty. Hence, the cultured cells may exhibit genetic heterogeneity from the very beginning. In contrast, plants raised from shoot apices have generally proved to be true- to-type.

ii. Cultured cells of several important plants do not exhibit totipotency. Moreover, in few cases these cells initially form organs/ embryos and whole plants but this potentiality is often lost after some time in culture. Besides, shoot apices possess a high regeneration ability which is retained in prolonged cultures. Shoot apices are mostly preferred to develop a virus free plants and also for the rapid clonal multiplication.

iii. Haploidy, which is highly unstable in callus and suspension cultures can be maintained through shoot tip culture and axillary-bud proliferation.

iv. The cells of shoot-tip and young embryos are small and meristematic. They ppear to be better suited than larger cells to survive liquid nitrogen (LN) freezing and thawing.

Pre-culture

In several cases, a brief culture of shoot apices for at least 48h at 4°C before freezing has proved beneficial for consistently high frequency of survival of shoot apices after freezing in liquid nitrogen. The other treatments include the application of additives that known to enhance plant stress tolerance, for example ABA, proline, osmoticum (sucrose, mannitol), dimethylsulfoide (DMSO, 1-5%). Sugars acts as osmotically effective agents, although they do not penetrate inside the cells. Dehydration of cells/tissues occurs in the presence of sugars during the preculture, which prevents lethal ice crystal formation during freezing. Proline may act by reducing the level of latent injury to the cells or it may actively participate in recovery metabolism.

Cryoprotective Treatments

There are two potential sources of cell damage during cryopreservation (1) Formation of large ice crystals, inside the cells, leading to rupture of organelle and the cell itself, (2) intracellular concentration of solutes increases to toxic levels before or during freezing as a result of dehydration. Addition of cryoprotectants controls the appearance of ice crystals in cells and protects these cells from the toxic solution effect. Cryoprotectants are categorized as: (a) Penetrating, which exert their protective colligative action, (b) Non-penetrating, which affect through osmotic dehydration. A large number of heterogeneous groups of compounds have been shown to possess cryoprotective properties with different efficiencies, e.g. glycerol, DMSO etc. Cryoprotectant depress both the freezing and super-cooling point of water, i.e. the temperature at which the homogeneous nucleation of ice occurs, thus, retarding the growth of ice crystal formation in cells and protect cells from toxic effect. The cryoprotectants used in cryopreservation are:

a. Alcohols: Ethylene glycol, glycerol, propylene glycol, sorbitol, mannitol.

b. Sulphur containing compounds: Amino acids, dimethyl sulphoxide (DMSO), sugar (glucose, saccharose).

c. Polymers: Hydroxyethyl amidon, polyethylene glycol, polyvinyl pyrrolidine.

Vitrification:

At a sufficiently low temperature, highly concentrated aqueous solutions of cryoprotective agents become so viscous that they solidify into an amorphous "glassy" state, without ice crystal formation (crystallization) at practical cooling rates, this phenomenon is called vitrification. The significance of vitrification in cryopreservation of biological materials is that the cells applied with highly concentrated solution of osmotially active compounds, are protected from internal damage from ice crystal formation during freezing. This pretreatment also causes dehydration of cells. The commonly used cryoprotectants are employed for vitrification like DMSO.

Cryoprotective dehydration:

If cells are sufficiently dehydrated they may be able to withstand immersion in liquid nitrogen without further application of traditional cryoprotectant mixtures. Dehydration can be achieved by growing the cultures in the presence of high concentration of osmotically active compounds (sugars) and /or air desiccations in laminar-air-flow cabinet or over silica gel. Dehydration reduces the amount of water available for the ice formation.

Encapsulation and dehydration:

This involves the encapsulation of tissues in calcium alginate beads which are pre-grown in liquid culture media containing high concentrations of sucrose. The beads are transferred to sterile airflow in a laminar cabinet and desiccated further. After these treatments, the cells are able to withstand exposure to liquid nitrogen without application of chemical cryoprotectants.

Freezing and Storage

The type of crystal water within stored cells is very important for survival of the tissue. Different tissues have different sensitivities for cooling rates. In general, there are three different types of freezing procedures:

Rapid freezing:

The plant material is placed in vials, liquid nitrogen is poured directly in the vial and dipping the vial into an open flask filled with liquid nitrogen. In this procedure, cooling between -10°C and -70°C occurred at the rate of >1000°C/min. The quicker the freezing is done, smaller the intracellular ice crystals are formed. In combination with desiccation or vitrification pre-treatments, ultra rapid cooling is proved to be the most attractive method for cryopreservation of plant materials. This method has been successfully used for the cryopreservation of shoot-tips, somatic embryos and embryonal axes from zygotic embryos of a number of plant species. The survival rate of cryopreserved tissues by this method is high and when the desiccation pretreatment is applied even the cryoprotectants are not required.

Slow freezing:

The tissue is slowly frozen at a slow cooling rate of 0.5-4°C/min from 0 to -100°C, and then transferred to liquid nitrogen. Survival of cells frozen at slow freezing rates may involve some beneficial effects of dehydration, which minimizes the amount of water that freezes intracellularly. Slow cooling permits the flow of water from the cells to the outside, thereby promoting extracellular ice formation instead of intracellular freezing. It is generally agreed that upon extracellular freezing the cytoplasm will be effectively concentrated and plant cells will survive better when adequately dehydrated. This has been successfully employed for cryopreservation of meristems of few plants and has proved especially successful with cells from suspension cultures.

Stepwise freezing:

Firstly, the material is cooled gradually (ca 1°C/min) or step-wise (5°C/min) to an optimum intermediate temperature (-30°C to -50°C) for about 30 min, and then rapidly cooled by dipping into liquid nitrogen. The method is highly favorable for freeze preservation of shoot apices and buds. It is equally successful to cells from suspension cultures.

The initial slow freezing reduces the amount of intracellular freezable water by dehydrating the cells. Early in the freezing process ice is formed first outside the cells, and the unfrozen protoplasm of cells loses water due to the vapor pressure deficit between the supercooled protoplasm and the external ice. This initial cooling, thus, acts as another pre-treatment for dehydration of the cells.

Storage

Maintaining the frozen material at the correct temperature is as important as proper freezing itself. Temperatures above -130°C may allow ice-crystal growth inside the cells and, as a result reduce their viability. Long-term storage of the material frozen at -196°C, therefore, requires a liquid nitrogen refrigerator.

Generally, the frozen cells or tissues are immediately kept for storage at temperature ranging from -70°C to -196°C. The storage is ideally done in liquid nitrogen refrigerator at -150°C in the vapor phase or -196°C in the liquid phase. The temperature should be sufficiently low for long term storage of cells to arrest all metabolic activities and to prevent biochemical injury.

Thawing

Rapid thawing of the material frozen at -196°C is achieved by plunging it into water at 37 to 40°C which gives thawing rate of 500-750°C/min. After about 90s, the material is transferred to an ice bath and maintained there until recultured or its viability is tested. The transfer is necessary because the cells might get damage if it is left long in the water bath 37-45°C. Rapid thawing protects the cells from the damaging effects of ice crystal formation, which may occur during slow warming.

Re-culturing

The material after thawing should be washed several times to remove the cryoprotectant which may otherwise be toxic to the cells. A gradual dilution of the cryoprotectant is desirable in-order to avoid any deplasmolysis injury to the cells. The plant material frozen at -196°C may need some special requirements for better survival when re-cultured. Shoot-tips from frozen seedlings of tomato directly developed into plantlets only if the medium was supplemented with GA_3. In its absence, apices callused, followed by the differentiation of adventitious shoots.

References

- Prestwich, G. D. Simplifying the extracellular matrix for 3-D cell culture and tissue engineering: A pragmatic approach. J. Cell. Biochem.101, 1370-1383, doi:10.1002/jcb.21386 (2007)
- Barsky, Eugene; Giustini, Dean (December 2007), "Introducing Web 2.0: wikis for health librarians" (PDF), Journal of the Canadian Health Libraries Association, 28 (4), pp. 147–150, doi:10.5596/c07-036, ISSN 1708-6892, retrieved November 7, 2011

- Souza, G. R. et al. Three-dimensional Tissue Culture Based on Magnetic Cell Levitation. Nature Nanotechnol.5, 291-296, doi:10.1038/nnano.2010.23 (2010)
- Goldman, Eric, "Wikipedia's Labor Squeeze and its Consequences", Journal on Telecommunications and High Technology Law, 8
- Majchrzak, A.; Wagner, C.; Yates, D. (2006), "Corporate wiki users: results of a survey", Proceedings of the 2006 international symposium on Wikis, Symposium on Wikis, pp. 99–104, doi:10.1145/1149453.1149472, ISBN 1-59593-413-8, retrieved April 25, 2011

4

Micropropagation and Embryogenesis of Plants

Micropropagation is the artificial process of producing plants vegetatively through tissue or cell culture techniques. The process by which plant embryos form and develop is called plant embryogenesis. The artificial replication of plant embryogenesis is known as somatic embryogenesis. This chapter has been carefully written to provide an easy understanding of micropropagation and plant embryogenesis.

Micropropagation

Micropropagation is the practice of rapidly multiplying stock plant material to produce a large number of progeny plants, using modern plant tissue culture methods.

Micropropagation is used to multiply noble plants such as those that have been genetically modified or bred through conventional plant breeding methods. It is also used to provide a sufficient number of plantlets for planting from a stock plant which does not produce seeds, or does not respond well to vegetative reproduction.

Cornell University botanist Frederick Campion Steward discovered and pioneered micropropagation and plant tissue culture in the late 1950s and early 1960s.

Stages of Micropropagation

Establishment

Micropropagation begins with the selection of plant material to be propagated. The plant tissues are removed from an intact plant in a sterile condition. Clean stock materials that are free of viruses and fungi are important in the production of the healthiest plants. Once the plant material is chosen for culture, the collection of explant(s) begins and is dependent on the type of tissue to be used; including stem tips, anthers, petals, pollen and other plant tissues. The explant material is then surface sterilized, usually in multiple courses of bleach and alcohol washes, and finally rinsed in sterilized water. This small portion of plant tissue, sometimes only a single cell, is placed on a growth medium, typically containing sucrose as an energy source and one or more plant growth regulators (plant hormones). Usually the medium is thickened with agar to create a gel which supports the explant during growth. Some plants are easily grown on simple media, but others require more complicated media for successful growth; the plant tissue grows and differentiates into new tissues depending on the medium. For example, media containing cytokinin are used to create branched shoots from plant buds.

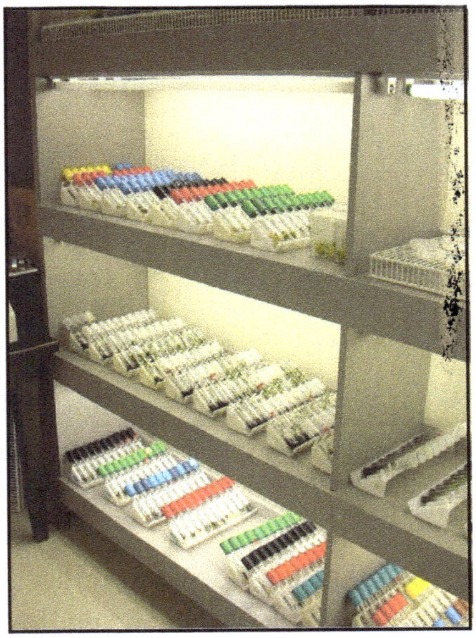

In vitro culture of plants in a controlled, sterile environment.

Multiplication

Multiplication is the taking of tissue samples produced during the first stage and increasing their number. Following the successful introduction and growth of plant tissue, the establishment stage is followed by multiplication. Through repeated cycles of this process, a single explant sample may be increased from one to hundreds and thousands of plants. Depending on the type of tissue grown, multiplication can involve different methods and media. If the plant material grown is callus tissue, it can be placed in a blender and cut into smaller pieces and recultured on the same type of culture medium to grow more callus tissue. If the tissue is grown as small plants called plantlets, hormones are often added that cause the plantlets to produce many small offshoots. After the formation of multiple shoots, these shoots are transferred to rooting medium with a high auxin\ cytokinin ratio. After the development of roots, plantlets can be used for hardening.

Pretransplant

"Hardening" refers to the preparation of the plants for a natural growth environment. Until this stage, the plantlets have been grown in "ideal" conditions, designed to encourage rapid growth. Due to the controlled nature of their maturation, the plantlets often do not have fully functional dermal coverings. This causes them to be highly susceptible to disease and inefficient in their use of water and energy. In vitro conditions are high in humidity, and plants grown under these conditions often do not form a working cuticle and stomata that keep the plant from drying out. When taken out of culture, the plantlets need time to adjust to more natural environmental conditions. Hardening typically involves slowly weaning the plantlets from a high-humidity, low light, warm environment to what would be considered a normal growth environment for the species in question.

Banana plantlets transferred to soil (with vermicompost) from plant media. This process is done for acclimatization of plantlets to the soil as they were previously grown in plant media. After growing for some days the plantlets are transferred to the field.

This stage involves treating the plantlets/shoots produced to encourage root growth and "hardening." It is performed *in vitro*, or in a sterile "test tube" environment.

Transfer from Culture

Plant tissue cultures being grown at a USDA seed bank, the National Center for Genetic Resources Preservation.

In the final stage of plant micropropagation, the plantlets are removed from the plant media and transferred to soil or (more commonly) potting compost for continued growth by conventional methods.

This stage is often combined with the "pretransplant" stage.

Methods of Micro-propagation

Meristem Culture

In Meristem culture the Meristem and a few subtending leaf primordial are placed into a suitable growing media. An elongated rooted platelet is produced after some weeks, and is transferred to the soil when it has attained a considerable height. A disease free plant can be produced by this method. Experimental result also suggest that this technique can be successfully utilized for rapid multiplication of various herbaceous Plants.

Callus Culture

A callus is mass of undifferentiated parenchymatous cells. When a living plant tissue is placed in an artificial growing medium with other conditions favorable, callus is formed. The growth of callus varies with the homogenous levels of auxin and Cyotkininn and can be manipulated by endogenous supply of these growth regulators in the culture medium. The callus growth and its organogenesis or embryogenesis can be referred into three different stages:

- Stage I: Rapid production of callus after placing the explants in culture medium.
- Stage II: The callus is transferred to other medium containing growth regulators for the induction of adventitious organs.
- Stage III: The new plantlet is then exposed gradually to the environmental condition.

Suspension Culture

A cell suspension culture refers to cells and or groups of cells dispersed and growing in an aerated liquid culture medium (Street, 1997, Thorpe1981) is placed in a liquid medium and shaken vigorously and balanced dose of hormones. Suezawa et al. (1988) reported Cyotkininn induced adventitious buds in kiwi fruit in a suspension culture sub- culture for about a week.

Embryo Culture

In embryo culture, the embryo is excised and placed into a culture medium with proper nutrient in aseptic condition. To obtain a quick and optimum growth into plantlets, it is transferred to soil. It is particularly important for the production of interspecific and intergeneric hybrids and to overcome the embryo abortion.

Protoplast Culture

In protoplast culture, the plant cell can be isolated with the help of wall degrading enzymes and growth in a suitable culture medium in a controlled condition for regeneration of plantlets. Under suitable conditions the protoplast develops a cell wall followed by an increase in cell division and differentiation and grows into a new plant. The protoplast are first cultured in liquid medium at 25 to 28 C with a light intensity of 100 to 500 lux or in dark and after undergoing substantial cell division, they are transferred into solid medium congenial or morphogenesis in many horticultural crops response well to protoplast culture.

Advantages

Micropropagation has a number of advantages over traditional plant propagation techniques:

- The main advantage of micropropagation is the production of many plants that are clones of each other.
- Micropropagation can be used to produce disease-free plants.
- It can have an extraordinarily high fecundity rate, producing thousands of propagules while conventional techniques might only produce a fraction of this number.
- It is the only viable method of regenerating genetically modified cells or cells after protoplast fusion.
- It is useful in multiplying plants which produce seeds in uneconomical amounts, or when plants are sterile and do not produce viable seeds or when seed cannot be stored.
- Micropropagation often produces more robust plants, leading to accelerated growth compared to similar plants produced by conventional methods - like seeds or cuttings.
- Some plants with very small seeds, including most orchids, are most reliably grown from seed in sterile culture.
- A greater number of plants can be produced per square meter and the propagules can be stored longer and in a smaller area.

Disadvantages

Micropropagation is not always the perfect means of multiplying plants. Conditions that limits its use include:

- It is very expensive, and can have a labour cost of more than 70%.
- A monoculture is produced after micropropagation, leading to a lack of overall disease resilience, as all progeny plants may be vulnerable to the same infections.
- An infected plant sample can produce infected progeny. This is uncommon as the stock plants are carefully screened and vetted to prevent culturing plants infected with virus or fungus.
- Not all plants can be successfully tissue cultured, often because the proper medium for growth is not known or the plants produce secondary metabolic chemicals that stunt or kill the explant.

- Sometimes plants or cultivars do not come true to type after being tissue cultured. This is often dependent on the type of explant material utilized during the initiation phase or the result of the age of the cell or propagule line.
- Some plants are very difficult to disinfect of fungal organisms.

The major limitation in the use of micropropagation for many plants is the cost of production; for many plants the use of seeds, which are normally disease free and produced in good numbers, readily produce plants in good numbers at a lower cost. For this reason, many plant breeders do not utilize micropropagation because the cost is prohibitive. Other breeders use it to produce stock plants that are then used for seed multiplication.

Mechanisation of the process could reduce labour costs, but has proven difficult to achieve, despite active attempts to develop technological solutions.

Adventitious Shoot Proliferation

Adventitious shoot proliferation in plant cell and tissue culture, in response to hormonal manipulation of the culture medium, require *de novo* differentiation of meristematic region, randomly, all over the tissue other than the pre-existing meristem. It is a multistep process and a series of intracellular events, collectively called induction that occurs before the appearance of morphologically recognizable organs. Micropropagation via adventitious shoot regeneration may occur directly or indirectly via an intervening callus phase. Indirect regeneration often results in somaclonal variations, making this strategy less desirable for large-scale clonal multiplication (Marcotrigiano and Jagannathan 1988; Thorpe et al. 1991). Therefore, regeneration of shoots directly from the explants is regarded as the most reliable method for clonal propagation. Various explants like leaf, cotyledon, embryo and root have been tried with different media combinations by the scientists to obtain adventitious shoot proliferation.

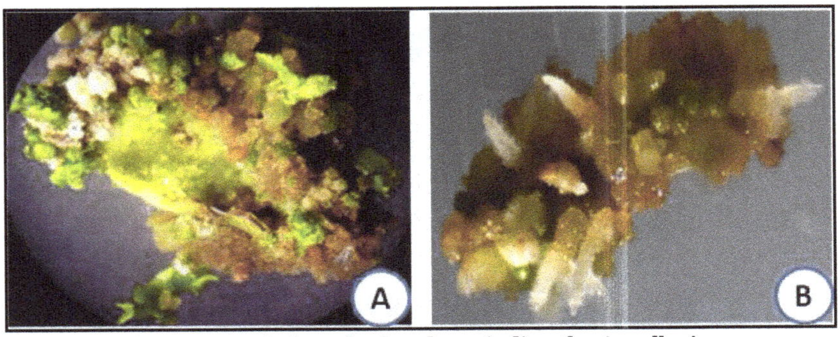

Organogenesis from leaf explants indirectly via callusing
A. Shoot differentiation B. Root differentiation.

Organogenic Differentiation

Regeneration of plant from the cultured explant may occur either through differentiation of shoot-buds or somatic embryogenesis.

Direct shoot proliferation from leaf-disc culture.

Direct differentiation of somatic embryos from hypocotyl explants.

The shoot-bud and embryo formation can be distinguished by the distinct morphological features. The shoot-bud is a monopolar structure. It develops from the procambial strands which establish a connection with the pre-existing vascular tissue dispersed within the callus or the cultured explants.

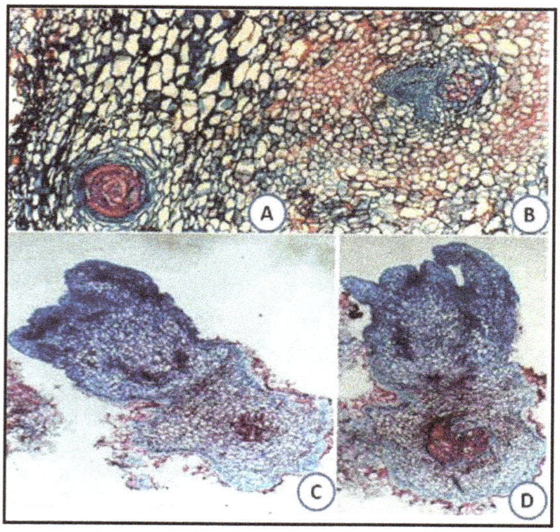

Shoot differentiation from callus tissue. A-B, Development of vascular nodules randomly in the callus, note a small shoot-bud originated from vascular tissue in figure B. C-D, Shoot-buds establish a connection with pre-existing vascular tissue developed from the callus.

Plant regeneration from isolated cells, protoplasts or unorganized mass of cells (callus) is generally more difficult than that obtained from the intact explants such as, cotyledons, hypocotyl segments and immature embryos. The regeneration obtained through de novo differentiation of shoot buds or somatic embryogenesis directly from explants may also exhibit genetic variability.

Induction of Organogenic Differentiation

Induction is a multistep process. A schematic representation of adventitious shoot proliferation from leaf-disc culture is presented. A series of intracellular events, collectively called induction, occur before the emergence of morphologically recognizable organs.

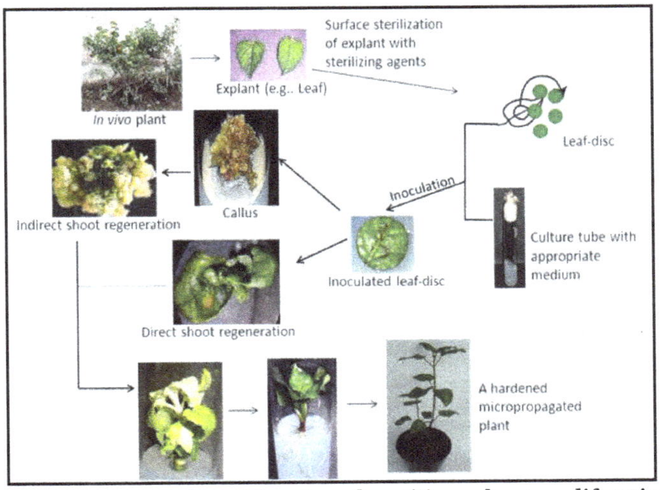

A schematic representation of *in vitro* adventitious shoot proliferation from leaf-disc culture.

Under the optimal growth regulator combinations, the cells induced to form a specific organ and would continue to develop into that organ even if the inductive growth regulators are removed. Hence, induction favours the irreversible commitment of cells to follow a particular developmental pathway. For example, *Brassica juncea,* undergoes the induction of organogenic differentiation where a cytokinin, BAP induces shoot-bud differentiation at the cut end of the cotyledon petiole. In the absence of BAP (basal medium) only roots are formed at the same site. The cotyledons transferred to basal medium after 11 days of incubation on BAP leads to the development of only shoots and no roots. Similarly, the cotyledons lose the potential to form shoots on BAP medium if they are pre-cultured on BAP free medium for more than 7 days.

Ontogeny of Shoot Buds

Under the optimal conditions, meristems formed from the callus are random and scattered. On transferring to the medium supporting organized growth promotes first the appearance of localized clusters of cambium like cells. These meristemoids (or nodules or growing centers), which may become vascularized due to the appearance of tracheidal cells in the centre, are the site for organ formation in the callus, as seen in above figure. Initially, the meristemoids exhibits plasticity and can form shoots and roots.

Factors Affecting Shoot-bud Differentiation

The genotype and plant growth regulators are well known to affect regeneration frequency. Plant growth regulators play a major role in the regeneration which mainly depends upon the concentration and type of growth regulators used. For *in vitro* differentiation genotype plays equally, if not more critical role as the growth regulator. Besides, there are certain other factors which play a critical role in regeneration are:

Explant

Regenerability of an explant is influenced by several factors such as, the organ from which it is derived, the physiological state of the explant like age of the explant, young vs. mature, position of the explant on the plant and the explant size. Orientation of the explant on the medium and the inoculation density may also affect shoot-bud differentiation. There may be a decline in the number of shoots per culture and the percent cultures showing regeneration with increasing age of the seedlings.

Preparation of Explant

Sharma et al (1990) studied that in cotyledon cultures of *Brassica juncea* , shoot buds or roots are formed at the cut end of the petiole, depending on the culture medium. Lamina lacks this potential. However, the presence of laminar tissue is essential for the petiolar cells to exhibit totipotency. Therefore, the ideal explant to achieve regeneration is the lamina together with a short (1mm) petiole.

Orientation of the Explant

Orientation of explant is proved to be critical for organogenic differentiation in cotyledon cultures of *B. juncea*. Inoculating the cotyledons with their abaxial surface (lower surface away from the stem) in contact with the medium and the petiolar cut end embedded in the medium gave best response. The explants in which due to expansion and curling of the lamina, the petiole lost contact with the medium within 3-5 days after culture, failed to form roots or shoots. Generally, the explants inoculated horizontally on the medium produced three times more shoots than those planted vertically.

Physical Factors

i. Explants grown on liquid or semi-solid medium give different degree of organogenesis. In few species, like tobacco, the medium with 1% agar showed only flower formation. With lowering the agar concentration the frequency of flower formation dropped and vegetative bud differentiation occurred. In liquid medium, the tissue exhibited callusing and vegetative bud formation.

ii. The quality of light also influences organogenic differentiation. Alternating light and dark period (diffused light, 15-16 hrs) proved best. Callus maintained under continuous light remained whitish and may not exhibit organogenesis. Blue light promote shoot-bud differentiation whereas red light stimulated rooting in tobacco. Calli of *Brassica oleracea* grown in dark for 20 days formed shoot-buds 12 days after transfer to light while those shifted to light after 12 days of growth in dark differentiated shoots within 9 days.

iii. Skoog (1944) studied the effect of a range of temperature on tobacco callus growth and differentiation. Growth of callus increased with rise in temperature up to 33°C, but for shoot-bud differentiation 18°C was optimum.

Axillary Shoot Proliferation

Axillary buds are usually present in the axil of each leaf and every bud has the potential to develop into a shoot. In nature these buds remain dormant for various periods. The species with strong apical dominance show the growth of axillary buds into shoot only if the terminal bud is removed or injured. The phenomenon of apical dominance is regulated by the interplay of growth regulators. The application of cytokinin to the axillary buds can overcome the apical dominance effect and stimulate the lateral buds to grow rapidly in the presence of terminal buds. If the exogenous growth regulator diminishes the lateral shoot stop growing.

Rate of Shoot Multiplication

In plant tissue culture, the rate of shoot multiplication can be determined by enhanced axillary branching. Due to continuous availability of cytokinin, the shoots formed by the bud, a priori present on the explant (nodal segment or shoot-tip cutting), develops axillary buds which may grow directly into shoots. This process may be repeated several times and the initial explant transformed into a mass of branches. There is a limit to which shoot multiplication can be achieved in a single passage, after which further axillary branching stops. At this stage, if shoots are excised and planted on a fresh medium of same composition, the shoot multiplication cycle can be repeated. This process can go on indefinitely, and can be maintained throughout the year independent of the season and the region.

In some plants, it may not be possible to break apical dominance by the application of growth regulator compositions, and the bud present a priori on the initial explant grows into an unbranched shoot. The rate of shoot multiplication in such cases would depend on the number of nodal cuttings that can be excised from the newly developed shoot at the end of each passage. With this alternative method of enhanced axillary branching, 6-7 fold shoot multiplication every 4-6 weeks could be achieved in the plants with strong apical dominance.

Factors Affecting Axillary Shoot Proliferation

Effects of season on culture establishment:

The extent of contamination as well as bud-break is highly dependent on the season. The cultures initiated during spring season (January to April) shows best response not only in terms of the frequency of bud-break but also in the vigor of the shoots with least contamination rate. Since, summer (May-August) is the period that concurs with rainy season in certain regions like India, the cultures are prone to infection. By winter the shoots become old and it is difficult to break the dormant state of the buds.

Effect of carbon source on shoot proliferation:

In cultured plant tissues, a continuous supply of carbohydrate from the medium is essential which are needed for growth and organized development of the plant and are necessary as a source of energy and carbon skeletons for biosynthetic process. For shoot induction from axillary buds, three carbon sources, sucrose, glucose and maltose are utilized in maximum plant tissue cultures at a fixed concentration of 30 g l^{-1} . Of these, sucrose is the most commonly used carbohydrate for plant tissue cultures and most culture media have it as the sole carbohydrate source. It favors higher growth of shoot, number of nodes per shoot and the rate of shoot multiplication compare to maltose and glucose. Sucrose is easily recognized and hydrolyzed by cell wall bound invertase into more efficiently utilizable forms of sugars, glucose and fructose which are incorporated into the cells. Glucose, derived from sucrose hydrolysis, is more accessible to the cultured tissues than glucose derived by maltose hydrolysis, due to a rapid sucrose hydrolysis but a slow maltose hydrolysis in the media.

Effect of growth regulators on shoot proliferation:

In general, cytokinins favors shoot proliferation and auxins favors root formation. In *S. acmella,* nodal explants bearing two opposite axillary buds were when cultured on MS basal medium or basal medium supplemented with BAP, Kinetin or 2-iP at 3 μM concentration, the frequency of bud-break was appreciable in basal medium but incorporation of BAP to the basal medium has further improved the incidence of bud-break and promoted multiple shoot formation (2 shoots/explant). While the least bud-break was observed on Kinetin supplemented medium and 2-iP was noticed to be inhibitory for axillary bud proliferation. The addition of a low concentration of GA_3 to the BAP supplemented medium further promoted multiple shoot formation. On the other hand, single shoot with long internodes was developed from axillary buds in cultures when NAA was added to BAP containing medium. The frequency of bud-break varied with the concentration of the BAP and at its optimum level of 5 μM, 10-fold shoot multiplication occurred every 5 weeks.

Axillary Shoot Proliferation Vs Adventitious Shoot Proliferation

i. The axillary shoot proliferation is the most popular approach to clonal propagation of crop plants because the cells of the shoot apex are uniformly diploid and are least susceptible to genotypic changes under culture conditions.

ii. Chimeras, whose breakdown is common during adventitious bud proliferation, are perpetuated in shoot-bud culture and, thus, the cause for change in ploidy sometimes. While axillary shoot proliferation favours genetically uniform plant formation.

iii. Moreover adventitious bud formation and callusing methods require denovo differentiation of shoot-buds which is not always possible.

iv. Further, the axillary shoot proliferation is comparatively a quicker method of shoot multiplication as pre-existing meristem only proliferate into shoots, thus, reducing the time required to form de novo meristem formation.

Axillary Shoot Proliferation Vs Conventional Method of Propagation

i. The conventional method of vegetative propagation by stem cuttings utilizes the ability of axillary buds to take over the function of main shoot in the absence of a terminal bud. However, the number of cuttings that can be taken from a selected plant in a year is extremely limited because in nature the vegetative growth is periodic. In *in vitro* conditions, axillary shoot proliferate irrespective of seasons and regions.

ii. A minimal size of cuttings required in conventional methods is around 24-30 cm in order to establish a plant from it. Thus, it may restrict the multiplication of plants if the stock of parent plant is limited or if the species is endangered.

iii. With axillary shoot proliferation, minimum cutting size required is <1cm, thus, it favours large scale multiplication even with the limited sample.

iv. With the axillary shoot proliferation method, juvenile nodal cuttings are made available throughout the year that helps to maintain faster rate of multiplication compare to conventional methods of vegetative propagation where juvenile phase is short lived and with mature cuttings it is difficult to establish propagation as the buds in the axil under go dormancy.

Experiment

Aim: Clonal propagation by axillary shoot proliferation from nodal explants of *Spilanthes acmella*.

Equipments: Autoclave, pH-meter, Magnetic stirrers, Magnetic beads, Weighing balance, Laminar-air-flow.

Materials Required : Salts and vitamins of Murashige and Skoog's (MS; 1962) , sucrose, agar, conical flasks, measuring cylinders and beakers of various sizes. Reagent glass bottles for storage, spatula, tissue rolls, distilled water. Cotton plugs, aluminium foils, muslin cloth, scissor, media stocks, 1N NaOH, 1N HCl, myo-inositol. Autoclavable polybags, rubber bands, borosil glass test-tubes (150mm x 25mm without rim). Black markers, micropippete, micropipette-tips, test-tube stands, autoclavable baskets, plastic trays, fresh stem cuttings of *S. acmella.*

Plant Growth regulators (Sigma):

Benzyl amino purine (BAP), 6-furfuryl amino purine (Kinetin), 6-γγ –Dimethylallyl-amino purine (2-iP), Naphthalene acetic acid (NAA), Indole-3-butyric acid.

Protocol

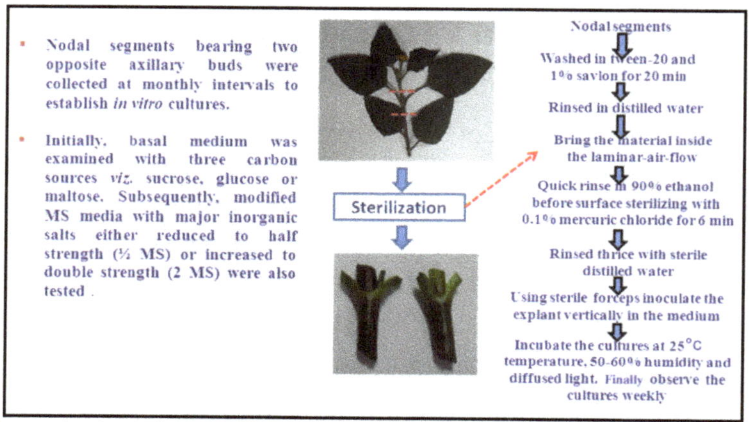

Schematic view of establishment of nodal segment cultures of *S. acmella* for clonal propagation.

Results

Shoot Multiplication

Nodal explants of *S. acmella* bearing two opposite axillary buds were cultured on MS basal medium or basal medium supplemented with BAP, Kinetin or 2-iP. MS + BAP medium proved optimum for shoot multiplication and results into 10-fold shoot multiplication at every 5 weeks. At the end of the passage each shoot was cut into single node segments and planted on the fresh medium of the same composition. Each node again produced multinodal, multiple shoots after 5 weeks. Shoot multiplication rate in various subculture (S) passages, S_1 =10.2, S_2 =10.3, S_3 =10.4, S_4 =10.6, S_5 =10.6, S_6 =10.5, S_7 =10.5, S_8 =10.5, S_9 =10.6, S_{10} =10.6, was >10 fold from S_1 to S_{10} on MS + BAP (5 μM). Since every time the explants were taken from freshly formed *in vitro* shoots, therefore, no significant difference (variation) was observed in the results.

Nodal segment culture: *invitro* clonal propagation of *Spilanthes acmella* by axillary shoots proliferation.

Rooting and Transplantation

Terminal 3-4 cm long portions of shoots after the growth period of 5 weeks are used for rooting. The remaining portions of the shoots can be cut into single node segments and utilized for further multiplication. MS basal medium was tested at full and half (½ MS) strengths of the major inorganic salts and the media are supplemented with 10, 30 and 50 gl^{-1} sucrose. Half MS was distinctly better than full MS basal medium, in terms of length of the shoot, percent rooting, number of roots per shoot and number of laterals present on the roots. The rooting was positively correlated with the sucrose concentration in the medium. On ½ MS + 50 gl^{-1} sucrose, shoots formed more than 35 roots directly from the basal cut end of the shoots. On this medium roots appeared after 2 weeks and maximum response was observed after 4 weeks. Some of the roots had developed laterals.

Rooted shoots were transferred out of culture. The plants were acclimatized by covering the pots with polythene bags to maintain high humidity for 6-7 days and irrigated with major salt solution of MS medium. After 7 days, 3-4 small holes were made in the bag and plantlets were irrigated as before but, at frequent intervals. After 25 days, polythene bags were removed and the acclimatized plants were transferred to a shaded area under natural conditions at a temperature range of 20-25°C with photoperiod of 12/12 h (light/dark). The plantlets were acclimatized successfully. During *in vitro* hardening, shoots elongated, leaves turned greener, and their lamina expanded. Consequently, the plants seemed much healthier and grew more vigorously after *in vitro* hardening.

Nodal segment culture: Rooting and acclimatization.

Meristem

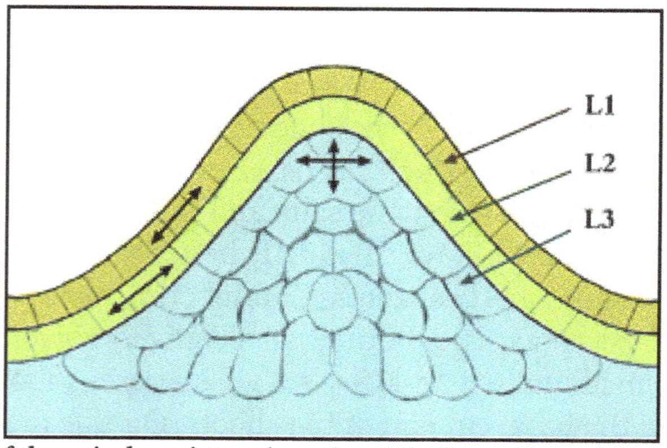

Tunica-Corpus model of the apical meristem (growing tip). The epidermal (L1) and subepidermal (L2) layers form the outer layers called the tunica. The inner L3 layer is called the corpus. Cells in the L1 and L2 layers divide in a sideways fashion, which keeps these layers distinct, whereas the L3 layer divides in a more random fashion.

A meristem is the tissue in most plants containing undifferentiated cells (meristematic cells), found in zones of the plant where growth can take place.

Meristematic cells give rise to various organs of the plant and keep the plant growing. The *shoot apical meristem* (SAM) gives rise to organs like the leaves and flowers, while the *root apical meristem* (RAM) provides the meristematic cells for the future root growth. SAM and RAM cells divide rapidly and are considered indeterminate, in that they do not possess any defined end status. In that sense, the meristematic cells are frequently compared to the stem cells in animals, which have an analogous behavior and function.

The term *meristem* was first used in 1858 by Karl Wilhelm von Nägeli (1817–1891) in his book *Beiträge zur Wissenschaftlichen Botanik* ("Contributions to Scientific Botany"). It is derived from the Greek word *merizein*, meaning to divide, in recognition of its inherent function.

In general, differentiated plant cells cannot divide or produce cells of a different type. Therefore, cell division in the meristem is required to provide new cells for expansion and differentiation of tissues and initiation of new organs, providing the basic structure of the plant body.

Meristematic cells are incompletely or not at all differentiated, and are capable of continued cellular division (youthful). Furthermore, the cells are small and protoplasm fills the cell completely. The vacuoles are extremely small. The cytoplasm does not contain differentiated plastids (chloroplasts or chromoplasts), although they are present in rudimentary form (proplastids). Meristematic cells are packed closely together without intercellular cavities. The cell wall is a very thin *primary cell wall.*

Maintenance of the cells requires a balance between two antagonistic processes: organ initiation and stem cell population renewal.

Apical meristems are the completely undifferentiated (indeterminate) meristems in a plant. These differentiate into three kinds of primary meristems. The primary meristems in turn produce the two secondary meristem types. These secondary meristems are also known as lateral meristems because they are involved in lateral growth.

At the meristem summit, there is a small group of slowly dividing cells, which is commonly called the central zone. Cells of this zone have a stem cell function and are essential for meristem maintenance. The proliferation and growth rates at the meristem summit usually differ considerably from those at the periphery.

Meristems also are induced in the roots of legumes such as soybean, *Lotus japonicus*, pea, and *Medicago truncatula* after infection with soil bacteria commonly called Rhizobium. Cells of the inner or outer cortex in the so-called "window of nodulation" just behind the developing root tip are induced to divide. The critical signal substance is the lipo-oligosaccharide Nod-factor, decorated with side groups to allow specificity of interaction. The Nod factor receptor proteins NFR1 and NFR5 were cloned from several legumes including *Lotus japonicus, Medicago truncatula* and soybean (*Glycine max*). Regulation of nodule meristems utilizes long distance regulation commonly called "Autoregulation of Nodulation" (AON). This process involves a leaf-vascular tissue located LRRreceptorkinases (LjHAR1, GmNARK and MtSUNN), CLE peptide signalling, and KAPP interaction, similar to that seen in the CLV1,2,3 system. LjKLAVIER also exhibits a nodule regulation phenotype though it is not yet known how this relates to the other AON receptor kinases.

Apical Meristems

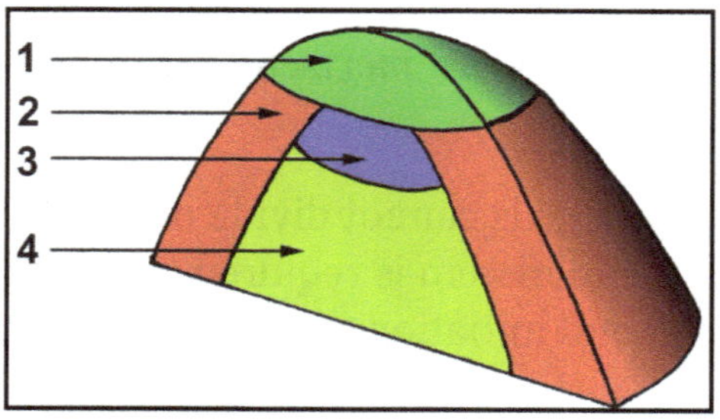

Organisation of an apical meristem (growing tip) 1 - Central zone 2 - Peripheral zone 3 - Medullary (i.e. central) meristem 4 - Medullary tissue.

The number of layers varies according to plant type. In general the outermost layer is called the tunica while the innermost layers are the corpus. In monocots, the tunica determine the physical characteristics of the leaf edge and margin. In dicots, layer two of the corpus determine the characteristics of the edge of the leaf. The corpus and tunica play a critical part of the plant physical appearance as all plant cells are formed from the meristems. Apical meristems are found in two locations: the root and the stem. Some Arctic plants have an apical meristem in the lower/middle parts of the plant. It is thought that this kind of meristem evolved because it is advantageous in Arctic conditions.

Shoot Apical Meristems

Shoot apical meristems of *Crassula ovata* (left). Fourteen days later, leaves have developed (right).

The source of all above-ground organs. Cells at the shoot apical meristem summit serve as stem cells to the surrounding peripheral region, where they proliferate rapidly and are incorporated into differentiating leaf or flower primordia.

The shoot apical meristem is the site of most of the embryogenesis in flowering plants. Primordia of leaves, sepals, petals, stamens and ovaries are initiated here at the rate of one every time interval, called a plastochron. It is where the first indications that flower development has been evoked are manifested. One of these indications might be the loss of apical dominance and the release of otherwise dormant cells to develop as auxiliary shoot meristems, in some species in axils of primordia as close as two or three away from the apical dome. The shoot apical meristem consists of 4 distinct cell groups:

- Stem cells.

- The immediate daughter cells of the stem cells.
- A subjacent organising centre.
- Founder cells for organ initiation in surrounding regions.

The four distinct zones mentioned above are maintained by a complex signalling pathway. In *Arabidopsis thaliana*, 3 interacting *CLAVATA* genes are required to regulate the size of the stem cell reservoir in the shoot apical meristem by controlling the rate of cell division. CLV1 and CLV2 are predicted to form a receptor complex (of the LRR receptor-like kinase family) to which CLV3 is a ligand. CLV3 shares some homology with the ESR proteins of maize, with a short 14 amino acid region being conserved between the proteins. Proteins that contain these conserved regions have been grouped into the CLE family of proteins.

CLV1 has been shown to interact with several cytoplasmic proteins that are most likely involved in downstream signalling. For example, the CLV complex has been found to be associated with Rho/Rac small GTPase-related proteins. These proteins may act as an intermediate between the CLV complex and a mitogen-activated protein kinase (MAPK), which is often involved in signalling cascades. KAPP is a kinase-associated protein phosphatase that has been shown to interact with CLV1. KAPP is thought to act as a negative regulator of CLV1 by dephosphorylating it.

Another important gene in plant meristem maintenance is *WUSCHEL* (shortened to *WUS*), which is a target of CLV signalling. *WUS* is expressed in the cells below the stem cells of the meristem and its presence prevents the differentiation of the stem cells. CLV1 acts to promote cellular differentiation by repressing *WUS* activity outside of the central zone containing the stem cells. *SHOOT MERISTEMLESS* (*STM*) also acts to prevent the differentiation of stem cells by repressing the expression of MYB genes that are involved in cellular differentiation.

Root Apical Meristem

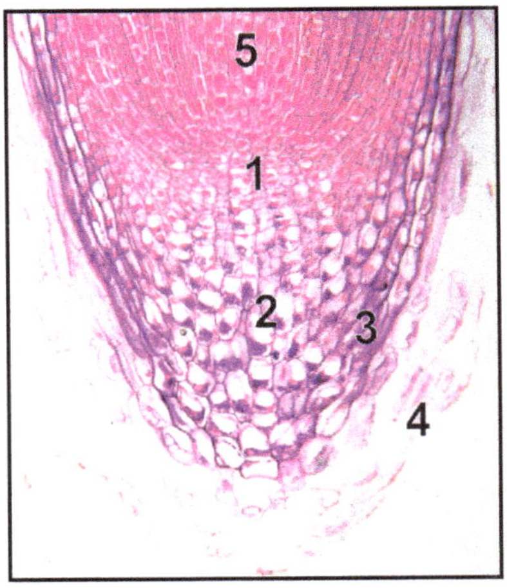

10x microscope image of root tip with meristem 1 - quiescent center 2 - calyptrogen (live rootcap cells) 3 - rootcap 4 - sloughed off dead rootcap cells 5 - procambium.

Unlike the shoot apical meristem, the root apical meristem produces cells in two dimensions. It harbors two pools of stem cells around an organizing center called the quiescent center (QC) cells and together produce most of the cells in an adult root. At its apex, the root meristem is covered by the root cap, which protects and guides its growth trajectory. Cells are continuously sloughed off the outer surface of the root cap. The QC cells are characterized by their low mitotic activity. Evidence suggests that the QC maintains the surrounding stem cells by preventing their differentiation, via signal(s) that are yet to be discovered. This allows a constant supply of new cells in the meristem required for continuous root growth. Recent findings indicate that QC can also act as a reservoir of stem cells to replenish whatever is lost or damaged. Root apical meristem and tissue patterns become established in the embryo in the case of the primary root, and in the new lateral root primordium in the case of secondary roots.

Intercalary Meristem

In angiosperms, intercalary meristems occur only in monocot (in particular, grass) stems at the base of nodes and leaf blades. Horsetails also exhibit intercalary growth. Intercalary meristems are capable of cell division, and they allow for rapid growth and regrowth of many monocots. Intercalary meristems at the nodes of bamboo allow for rapid stem elongation, while those at the base of most grass leaf blades allow damaged leaves to rapidly regrow. This leaf regrowth in grasses evolved in response to damage by grazing herbivores, but is more familiar to us in response to lawnmowers.

Floral Meristem

When plants begin the developmental process known as flowering, the shoot apical meristem is transformed into an inflorescence meristem, which goes on to produce the floral meristem, which produces the sepals, petals, stamens, and carpels of the flower.

In contrast to vegetative apical meristems and some exflorescence meristems, floral meristems cannot continue to grow indefinitely. Their future growth is limited to the flower with a particular size and form. The transition from shoot meristem to floral meristem requires floral meristem identity genes, that both specify the floral organs and cause the termination of the production of stem cells. *AGAMOUS* (*AG*) is a floral homeotic gene required for floral meristem termination and necessary for proper development of the stamens and carpels. *AG* is necessary to prevent the conversion of floral meristems to inflorescence shoot meristems, but is not involved in the transition from shoot to floral meristem. AG is turned on by the floral meristem identity gene *LEAFY* (*LFY*) and *WUS* and is restricted to the centre of the floral meristem or the inner two whorls. This way floral identity and region specificity is achieved. WUS activates AG by binding to a consensus sequence in the AG's second intron and LFY binds to adjacent recognition sites. Once AG is activated it represses expression of WUS leading to the termination of the meristem.

Through the years, scientists have manipulated floral meristems for economic reasons. An example is the mutant tobacco plant "Maryland Mammoth." In 1936, the department of agriculture of Switzerland performed several scientific tests with this plant. "Maryland Mammoth" is peculiar in that it grows much faster than other tobacco plants.

Apical Dominance

Apical dominance is phenomenon where one meristem prevents or inhibits the growth of other meristems. As a result, the plant will have one clearly defined main trunk. For example, in trees, the tip of the main trunk bears the dominant meristem. Therefore, the tip of the trunk grows rapidly and is not shadowed by branches. If the dominant meristem is cut off, one or more branch tips will assume dominance. The branch will start growing faster and the new growth will be vertical. Over the years, the branch may begin to look more and more like an extension of the main trunk. Often several branches will exhibit this behaviour after the removal of apical meristem, leading to a bushy growth.

The mechanism of apical dominance is based on the plant hormone auxin. It is produced in the apical meristem and transported towards the roots in the cambium. If apical dominance is complete, it prevents any branches from forming as long as the apical meristem is active. If the dominance is incomplete, side branches will develop.

Recent investigations into apical dominance and the control of branching have revealed a new plant hormone family termed strigolactones. These compounds were previously known to be involved in seed germination and communication with mycorrhizal fungi and are now shown to be involved in inhibition of branching.

Diversity in Meristem Architectures

Is the mechanism of being *indeterminate* conserved in the SAMs of the plant world? The SAM contains a population of stem cells that also produce the lateral meristems while the stem elongates. It turns out that the mechanism of regulation of the stem cell number might indeed be evolutionarily conserved. The *CLAVATA* gene *CLV2* responsible for maintaining the stem cell population in *Arabidopsis thaliana* is very closely related to the Maize gene *FASCIATED EAR 2*(*FEA2*) also involved in the same function. Similarly, in Rice, the *FON1-FON2* system seems to bear a close relationship with the CLV signaling system in *Arabidopsis thaliana*. These studies suggest that the regulation of stem cell number, identity and differentiation might be an evolutionarily conserved mechanism in monocots, if not in angiosperms. Rice also contains another genetic system distinct from *FON1-FON2*, that is involved in regulating stem cell number. This example underlines the innovation that goes about in the living world all the time.

Role of the KNOX-family Genes

Note the long spur of the above flower. Spurs attract pollinators and confer pollinator specificity. *(Flower:Linaria dalmatica).*

Complex leaves of *C. hirsuta* are a result of KNOX gene expression.

Genetic screens have identified genes belonging to the KNOX family in this function. These genes essentially maintain the stem cells in an undifferentiated state. The KNOX family has undergone quite a bit of evolutionary diversification, while keeping the overall mechanism more or less similar. Members of the KNOX family have been found in plants as diverse as Arabidopsis thaliana, rice, barley and tomato. KNOX-like genes are also present in some algae, mosses, ferns and gymnosperms. Misexpression of these genes leads to formation of interesting morphological features. For example, among members of *Antirrhinae*, only the species of genus Antirrhinum lack a structure called spur in the floral region. A spur is considered an evolutionary innovation because it defines pollinator specificity and attraction. Researchers carried out transposon mutagenesis in *Antirrhinum majus*, and saw that some insertions led to formation of spurs that were very similar to the other members of *Antirrhinae*, indicating that the loss of spur in wild *Antirrhinum majus* populations could probably be an evolutionary innovation.

The KNOX family has also been implicated in leaf shape evolution. One study looked at the pattern of KNOX gene expression in *A. thaliana*, that has simple leaves and *Cardamine hirsuta*, a plant having complex leaves. In *A. thaliana*, the KNOX genes are completely turned off in leaves, but in *C.hirsuta*, the expression continued, generating complex leaves. Also, it has been proposed that the mechanism of KNOX gene action is conserved across all vascular plants, because there is a tight correlation between KNOX expression and a complex leaf morphology.

Primary Meristems

Apical meristems may differentiate into three kinds of primary meristem:

- Protoderm: lies around the outside of the stem and develops into the epidermis.
- Procambium: lies just inside of the protoderm and develops into primary xylem and primary phloem. It also produces the vascular cambium, and cork cambium, secondary meristems. The cork cambium further differentiates into the phelloderm (to the inside) and the phellem, or cork (to the outside). All three of these layers (cork cambium, phellem and phelloderm) constitute the periderm. In roots, the procambium can also give rise to the pericycle, which produces lateral roots in eudicots.
- Ground meristem: develops into the cortex and the pith. Composed of parenchyma, collenchyma and sclerenchyma cells.

These meristems are responsible for primary growth, or an increase in length or height, which were discovered by scientist Joseph D. Carr of North Carolina in 1943.

Secondary Meristems

There are two types of secondary meristems, these are also called the *lateral meristems* because they surround the established stem of a plant and cause it to grow laterally (i.e., larger in diameter).

- Vascular cambium, which produces secondary xylem and secondary phloem. This is a process that may continue throughout the life of the plant. This is what gives rise to wood in plants. Such plants are called arborescent. This does not occur in plants that do not go through secondary growth (known as herbaceous plants).
- Cork cambium, which gives rise to the periderm, which replaces the epidermis.

Indeterminate Growth of Meristems

Though each plant grows according to a certain set of rules, each new root and shoot meristem can go on growing for as long as it is alive. In many plants, meristematic growth is potentially indeterminate, making the overall shape of the plant not determinate in advance. This is the primary growth. Primary growth leads to lengthening of the plant body and organ formation. All plant organs arise ultimately from cell divisions in the apical meristems, followed by cell expansion and differentiation. Primary growth gives rise to the apical part of many plants.

The growth of nitrogen fixing nodules on legume plants such as soybean and pea is either determinate or indeterminate. Thus, soybean (or bean and Lotus japonicus) produce determinate nodules (spherical), with a branched vascular system surrounding the central infected zone. Often, Rhizobium infected cells have only small vacuoles. In contrast, nodules on pea, clovers, and Medicago truncatula are indeterminate, to maintain (at least for some time) an active meristem that yields new cells for Rhizobium infection. Thus zones of maturity exist in the nodule. Infected cells usually possess a large vacuole. The plant vascular system is branched and peripheral.

Cloning

Under appropriate conditions, each shoot meristem can develop into a complete new plant or clone. Such new plants can be grown from shoot cuttings that contain an apical meristem. Root apical meristems are not readily cloned, however. This cloning is called asexual reproduction or vegetative reproduction and is widely practiced in horticulture to mass-produce plants of a desirable genotype. This process is also known as mericloning.

Propagating through cuttings is another form of vegetative propagation that initiates root or shoot production from secondary meristematic cambial cells. This explains why basal 'wounding' of shoot-borne cuttings often aids root formation.

Meristem Culture For Virus Elimination

The plants infected with bacteria and fungi can be treated by bactericidal and fungicidal compounds, there is no commercially available treatment to cure virus-infected plants. A large numbers of viruses are not transmitted through seeds. Therefore, it would be possible to obtain virus free plants from infected individuals by using seeds as propagules. However, genetic variation often occurs from the sexually reproduced plants when propagated by seeds. Generally, clonal multiplication of cultivars can be achieved by vegetative propagation. However, where the entire population of the clone is infected the only way to obtain pathogen-free stock is to eradicate the pathogen from vegetative parts of the plants and regenerate full plants from such tissues. Once pathogen free plants are obtained they can be multiplied indefinitely under conditions which would protect them from chance reinfection.

In Vitro Meristem and Shoot-tip Culture

Explant Terminology

The scientist who have attempted to recover pathogen-free plants through tissue culture techniques have indiscriminately designated the explants required to initiate cultures as 'shoot-tip', 'tip-meristem' and 'meristem-tip'. The apical meristem of a shoot is the portion lying distal to the youngest leaf primordium, it measures up to about 100μm in diameter and 250μm in length. The apical meristem together with one to three young leaf primordia, measuring 100-500μm, constitutes the shoot-apex. Although the chances of eradicating viruses is higher through 'meristem' culture, in most successful reports virus-free plants have been raised by culturing 100-1000μm long explants which could be according to the above definition is referred as 'shoot-tip'. To distinguish it from the *in vivo* technique of propagation through shoot-tip cuttings, the term 'meristem-tip' culture has been preferred for *in vitro* culture of small shoot-tips.

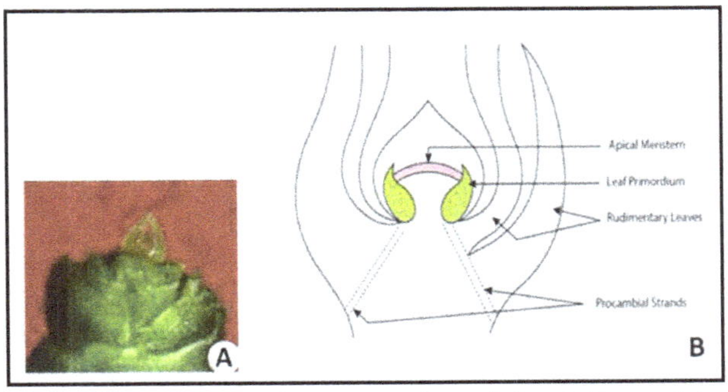

A shoot-tip explant; **B.** A cross section of shoot-tip meristem.

Reasons for Escape of Meristem from Virus

It is well known that the distribution of viruses in plants is uneven. In infected plants the apical meristems are generally either free or carry a very low concentration of the viruses. In older tissues the virus titer increases with increasing distance from the meristem-tips. The reasons proposed for the escape of meristem from virus invasion are:

(a) Viruses readily move in a plant body through the vascular system which is absent in the meristem.

(b) The alternative method of cell-to-cell movement of the virus through plasmodesmata is rather too slow to keep pace with the actively growing tip.

(c) High metabolic activity in the actively dividing meristem cells does not allow virus replication.

(d) The 'virus inactivating systems' in the plant body, if any, has higher activity in the meristem than in any other region. Thus, the meristem is protected from infection.

(e) A high endogenous auxin level in shoot apices may inhibit virus multiplication.

Factors Affecting Eradication of Virus Through Meristem Tip Culture

Culture medium, explant size and incubation conditions affecting plant regeneration from meristem-tip cultures have pronounced effect on virus eradication. Besides, thermotherapy or chemotherapy and physiological stage of the explants also affect virus elimination by shoot-tip culture.

Culture Medium

The nutrients, growth regulators and nature of the medium highly influence the development of virus free plants from meristem tip cultures. Maximum success is achieved from Murashige & Skoog's (MS) medium which promoted healthy, green shoot development compare to other nutrient media. The main reason for the suitability of medium for meristem-tip culture could be the presence of high levels of K^+ and NH_4^+ ions. There is no critical assessment on the role of various vitamins or amino acids but sucrose or glucose is the most commonly used carbon source in the medium, at the range of 2-4%, to raise virus free plants from meristem-tip cultures.

Large meristem-tip explants, measuring 500µm or more in length, may give rise to plants even in the basal medium but generally the presence of an auxin or a cytokinin or both plays a major role in the development of excised apical meristem. In angiosperms, the meristematic dome in the shoot-tip does not synthesize auxin on its own, but it is supplied by the second pair of youngest leaf primordia. Therefore, for development of excised meristem in culture, without the leaf primordia, requires the supply of exogenous auxin. The plants requiring only auxin must have a high endogenous cytokinin level in their meristems. Among auxins, the use of 2,4-D should be avoided which promotes only callusing. NAA and IAA are widely used auxins and NAA being preferred due to better stability. The role of GA_3 is also emphasized by few authors which is suggested to promote better growth and differentiation and suppresses callusing from meristem explants. Both liquid and semi-solid media have been tried for meristem–tip culture but, agar medium is generally preferred.

Explant Size

The survival of the meristem tips, under the controlled condition, is determined by the size of the explant. The larger the explant, the greater are the chances of plant regeneration. However, the survival of the explants can not be treated independent of the efficiency with which virus elimination is achieved that is inversely related to the size of the explant. Thus, explants should be small enough to eradicate viruses and large enough to be able to develop into a complete plant. Besides the size of the explant, the presence of leaf primordia influences the ability of the meristems to form plants. In some plants it is essential to excise shoot meristems with two to three leaf primordia. Smith and Murashige (1970) have suggested that leaf primordia supply auxin and cytokinin to the meristem necessary for its growth and differentiation. In a culture medium containing essential growth regulators, the excised meristems domes develop bipolar axes very quickly during reorganization. Once the root-shoot axis is established further development follows the same pattern as that of seedlings.

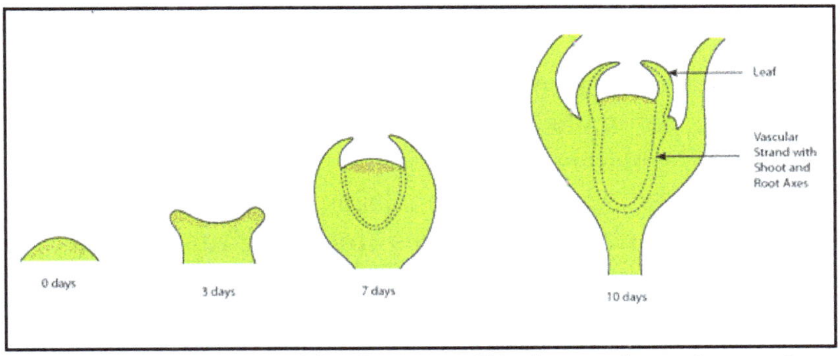

Schematic representation of development of bipolar axes by meristem culture.

Storage Conditions

Generally, light incubation of meristem tip culture is found better than dark incubation. The light intensity could range from 100 lx to 4000 lx which should increase in succession as the differentiation of meristem explant progresses. There are no clear information on the effect of temperature on regeneration of plants from excised meristem tips. The cultures are normally stored under room temperature (25±2°C) conditions.

Physiological Conditions of the Explant

Meristem-tips should be collected from actively growing buds. In few cases, the tips taken from terminal buds proved better than those taken from axillary buds. Seeing the higher number of axillary buds present per shoot, in majority of the reports axillary buds were utilized as explants to increase the overall production of virus-free plants. The time excision of buds is also critical, specially for the trees with periodic growth. For example, in temperate trees the growth of the plant is limited to only a very short period in the spring and afterwards dormancy starts. In such cases, the meristem-tip cultures can be raised during spring only for increased success rate.

Thermotherapy

Often, apical meristems are not always free of virus and it can't be considered as a universal occurrence. There are certain viruses like, Tobacco Mosaic Virus (TMV), Potato Virus X (PVX) and Cucumber Mosaic Virus (CMV), which invade the meristematic region of the growing tips and interrupts the growth of the meristematic tissue. In such cases also it has been possible to obtain virus-free plants by combining meristem-tip culture with thermotherapy. In this technique, first the mother plants are exposed to heat treatment before excising the meristem-tips or, alternatively, shoot-tip cultures are exposed to high temperature regimes (35°C-40°C) for certain duration (6h to 6 weeks) to obtain virus free plants. In the later case, continuous exposure to very high temperature causes deterioration of the host tissues. The first procedure of treating the mother plant has added advantage where larger explants can be taken from the treated stock and thus, favors relatively higher chances of the tip survival.

Chemotherapy

Chemotherapy is the treatment of an ailment by chemicals especially by killing micro-organisms. It will not eradicate the virus completely. However, a large number of antibiotics, growth regulators, amino acids, purines and pyrimidines can be tested for inactivation of viruses. A nucleotide analogue ribavirin has been found to be the most efficient viracide for plant viruses. This broad spectrum antiviral agent, effective against both plant and animal, was reported to eliminate PVY, CMV and TMV from tobacco explant cultures, Chlorotic Leaf Spot Virus (CLSV) in apple cultures when incorporated into the medium. Vidarabine (adenine arabinoside) and antiserum are also known to reduce the titre of viruses. The effectivity of the compound may vary with the virus and the host genotype.

Virus Elimination Through Callus Culture

It is a general observation that not all the cells in a calli uniformly carry the pathogen when raised from infected tissues. The two possible reasons for the escape of some cells of a systematically infected callus from virus infection are: (a) virus replication is unable to keep pace with cell proliferation, and (b) some cells acquire resistance to virus infection through mutagenesis. Therefore, it is possible to raise virus-free plants from infected shoot-tip calli. However, genetic instability of cultured cells and lack of plant regeneration in callus cultures of some plants poses the limitations of using calli for virus elimination.

Virus Indexing

Even after subjecting the meristem-tips to various treatments favoring virus eradication, only a proportion of the cultures yield virus free plants. Therefore, it is required to test all plants, regenerated through meristem-tip or callus cultures, for specific viruses before being used as mother plant to produce virus-free stock. The individual plants consistently showing negative results for virus titre can be marked as 'virus tested' for specific virus/es and can be released for commercial purposes. The following tests can be performed for virus testing:

i. The simplest test for the presence or absence of viruses in plant tissues is to examine the leaves and stem for the visible symptoms characteristic of the virus.

ii. Another test is the sap transmission test or 'bioassay test' or 'infectivity test'. It is a very sensitive test and can be performed at a commercial scale. To perform this, ground the test leaves in equal volume (w/v) of 0.5M phosphate buffer using a mortar and pestle. Leaves of the indicator plant (a plant very susceptible to specific viruses), dusted with 600-grade carborundum, are swabbed with the leaf sap from the test plant. After 5 min the incubated leaves are gently washed with water to remove the residual inoculum. The inoculated indicator plants are maintained in a glasshouse, separate from other plants. It may take several days to several weeks, depending on the nature of virus and the virus titre, for the symptoms to appear on the indicator plants. It is used to detect some viruses and viroids but is a slow process requiring several days to months.

iii. The third method, enzyme-linked immunosorbant assay (ELISA), is more rapid serological test which allows quick detection of important viruses. It relies on the use of antibodies prepared against the viral coat protein, requires only a small amount of antiserum and can be performed with simple equipment. However, it is not applicable to viroids and viruses which have lost their coat proteins.

Plant Embryogenesis

Plant embryogenesis is the process that produces a plant embryo from a fertilized ovule by asymmetric cell division and the differentiation of undifferentiated cells into tissues and organs. It occurs during seed development, when the single-celled zygote undergoes a programmed pattern of cell division resulting in a mature embryo. A similar process continues during the plant's life within the meristems of the stems and roots.

Seeds

Embryogenesis occurs naturally as a result of sexual fertilization and the formation of the *zygotic embryos*. The embryo along with other cells from the mother plant develops into the seed or the next generation, which, after germination, grows into a new plant.

Embryogenesis may be divided up into two phases, the first involves morphogenetic events which form the basic cellular pattern for the development of the shoot-root body and the primary tissue layers; it also programs the regions of meristematic tissue formation. The second phase, or postembryonic development, involves the maturation of cells, which involves cell growth and the storage of macromolecules (such as oils, starches and proteins) required as a 'food and energy supply' during germination and seedling growth. Embryogenesis involves cell growth and division, cell differentiation and programmed cellular death. The zygotic embryo is formed following double fertilisation of the ovule, giving rise to two distinct structures: the plant embryo and the endosperm which together go on to develop into a seed. Seeds may also develop without fertilization, which is referred to as apomixis. Plant cells can also be induced to form embryos in plant tissue culture; such embryos are called *somatic embryos* and can be used to generate new plants from single cells.

Following fertilization, the zygote undergoes an asymmetrical cell division that gives rise to a small apical cell, which becomes the embryo and a large basal cell (called the suspensor), which functions to provide nutrients from the endosperm to the growing embryo, and the product is an ovoid mass of tissue called the proembryo. From the eight cell stage (octant stage) onwards, the zygotic embryo shows clear embryo patterning, which forms the main axis of polarity, and the linear formation of future structures. These structures include the shoot meristem, cotyledons, hypocotyl, and the root and root meristem: they arise from specific groups of cells as the young embryo divides and their formation has been shown to be position-dependent.

In the globular stage, the embryo develops radial patterning through a series of cell divisions, with the outer layer of cells differentiating into the 'protoderm.' The globular embryo can be thought of as two layers of inner cells with distinct developmental fates; the apical layer will go on to produce cotyledons and shoot meristem, while the lower layer produces the hypocotyl and root meristem. Bilateral symmetry is apparent from the heart stage; provascular cells will also differentiate at this stage. In the subsequent torpedo and cotyledonary stages of embryogenesis, the embryo completes its growth by elongating and enlarging.

In a dicot embryo, the hypophysis, which is the uppermost cell of the suspensor, differentiates to form part of the root cap called Columella.

Plant Growth and Buds

Embryonic tissue is made up of actively growing cells and the term is normally used to describe the early formation of tissue in the first stages of growth. It can refer to different stages of the sporophyte and gametophyte plant; including the growth of embryos in seedlings, and to meristematic tissues, which are in a persistently embryonic state, to the growth of new buds on stems.

In both gymnosperms and angiosperms, the young plant contained in the seed, begins as a developing egg-cell formed after fertilization (sometimes without fertilization in a process called apomixis) and becomes a plant embryo. This embryonic condition also occurs in the buds that form on stems. The buds have tissue that has differentiated but not grown into complete structures. They can be in a resting state, lying dormant over winter or when conditions are dry, and then commence growth when conditions become suitable. Before they start growing into stem, leaves, or flowers, the buds are said to be in an embryonic state.

Somatic Embryogenesis

Somatic embryos are formed from plant cells that are not normally involved in the development of embryos, i.e. ordinary plant tissue. No endosperm or seed coat is formed around a somatic embryo. Applications of this process include: clonal propagation of genetically uniform plant material; elimination of viruses; provision of source tissue for genetic transformation; generation of whole plants from single cells called protoplasts; development of synthetic seed technology. Cells derived from competent source tissue are cultured to form an undifferentiated mass of cells called a callus. Plant growth regulators in the tissue culture medium can be manipulated to induce callus formation and subsequently changed to induce embryos to form the callus. The ratio of different plant growth regulators required to induce callus or embryo formation varies with the type of plant. Asymmetrical cell division also seems to be important in the development of somatic embryos, and while failure to form the suspensor cell is lethal to zygotic embryos, it is not lethal for somatic embryos.

Somatic Embryogenesis

Somatic embryogenesis is an artificial process in which a plant or embryo is derived from a single somatic cell or group of somatic cells. Somatic embryos are formed from plant cells that are not normally involved in the development of embryos, i.e. ordinary plant tissue. No endosperm or seed coat is formed around a somatic embryo. Applications of this process include: clonal propagation of genetically uniform plant material; elimination of viruses; provision of source tissue for genetic transformation; generation of whole plants from single cells called protoplasts; development of synthetic seed technology. Cells derived from competent source tissue are cultured to form an undifferentiated mass of cells called a callus. Plant growth regulators in the tissue culture medium can be manipulated to induce callus formation and subsequently changed to induce embryos to form from the callus. The ratio of different plant growth regulators required to induce callus or embryo formation varies with the type of plant. Somatic embryos are mainly produced *in vitro* and for laboratory purposes, using either solid or liquid nutrient media which contain plant growth regulators (PGR's). The main PGRs used are auxins but can contain cytokinin in a smaller amount. Shoots and roots are monopolar while somatic embryos are bipolar, allowing them to form a whole plant without culturing on multiple media types. Somatic embryogenesis has served as a model to understand the physiological and biochemical events that occur during plant developmental processes as well as a component to biotechnological advancement. The first documentation of somatic embryogenesis was by Steward et al. in 1958 and Reinert in 1959 with carrot cell suspension cultures.

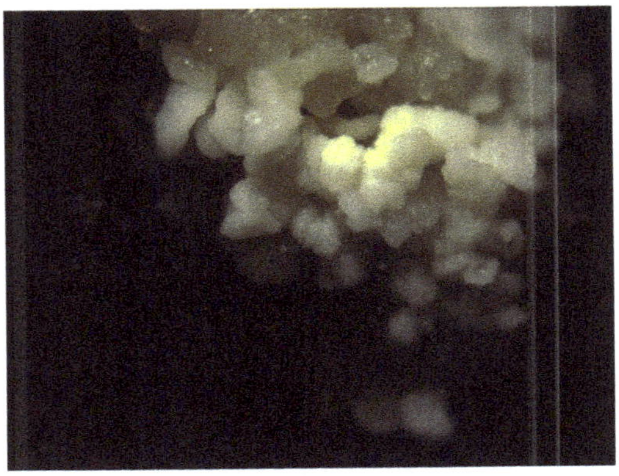

Switchgrass somatic embryos.

Direct and Indirect Embryogenesis

Somatic embryogenesis has been described to occur in two ways: directly or indirectly. Direct embryogenesis occurs when embryos are started directly from explant tissue creating an identical clone. Indirect embryogenesis occurs when explants produced undifferentiated, or partially differentiated, cells (often referred to as callus) which then is maintained or differentiated into plant tissues such as leaf, stem, or roots.

Plant Regeneration by Somatic Embryogenesis

Plant regeneration via somatic embryogenesis occurs in five steps: initiation of embryogenic cultures, proliferation of embryogenic cultures, prematuration of somatic embryos, maturation of somatic embryos and plant development on nonspecific media. Initiation and proliferation occur on a medium rich in auxin, which induces differentiation of localized meristematic cells. The auxin typically used is 2,4-D. Once transferred to a medium with low or no auxin, these cells can then develop into mature embryos. Germination of the somatic embryo can only occur when it is mature enough to have functional root and shoot apices.

Factors influencing Somatic Embryogenesis

Factors and mechanisms controlling cell differentiation in somatic embryos are relatively ambiguous. Certain compounds excreted by plant tissue cultures and found in culture media have been shown necessary to coordinate cell division and morphological changes. These compounds have been identified by Chung et al. as various polysaccharides, amino acids, growth regulators, vitamins, low molecular weight compounds and polypeptides. Several signaling molecules known to influence or control the formation of somatic embryos have been found and include extracellular proteins, arabinogalactan proteins and lipochitooligosaccharides. Temperature and lighting can also affect the maturation of the somatic embryo.

Uses of Somatic Embryogenesis

- Plant transformations.
- Mass propagation.

Forestry Related Example

The development of somatic embryogenesis procedures has given rise to research on seed storage proteins (SSPs) of woody plants for tree species of commercial importance, i.e., mainly gymnosperms, including white spruce. In this area of study, SSPs are used as markers to determine the embryogenic potential and competency of the embryogenic system to produce a somatic embryo biochemically similar to its zygotic counterpart (Flinn et al. 1991, Beardmore et al. 1997).

Grossnickle et al. (1992) compared interior spruce seedlings with emblings during nursery development and through a stock quality assessment program immediately before field outplanting. Seedling shoot height, root collar diameter, and dry weight increased at a greater rate in seedlings than in emblings during the first half of the first growing season, but thereafter shoot growth was similar among all plants. By the end of the growing season, seedlings were 70% taller than emblings, had greater root collar diameter, and greater shoot dry weight. Root dry weight increased more rapidly in seedlings than in emblings during the early growing season.

During fall acclimation, the pattern of increasing dormancy release index and increasing tolerance to freezing was similar in both seedlings and emblings. Root growth capacity decreased then increased during fall acclimation, with the increase being greater in seedlings.

Assessment of stock quality just prior to planting showed that: emblings had greater water use efficiency with decreasing predawn shoot water potential compared with seedlings; seedlings and emblings had similar water movement capability at both high and low root temperatures; net photosynthesis and needle conductance at low root temperatures were greater in seedlings than in emblings; and seedlings had greater root growth than emblings at 22°C root, but root growth among all plants was low at 7.5°C root temperature.

Growth and survival of interior spruce 313B Styroblock® seedlings and emblings after outplanting on a reforestation site were determined by Grossnickle and Major (1992). For both seedlings and emblings, osmotic potential at saturation (ψ_{sat}) and turgor loss point (ψ_{tip}) increased from a low of -1.82 and -2.22 MPa, respectively, just prior to planting to a seasonal high of -1.09 and -1.21 MPa, respectively, during active shoot elongation. Thereafter, seedlings and emblings (ψ_{sat}) and (ψ_{tip}) declined to -2.00 and -2.45 MPa, respectively, at the end of the growing season, which coincided with the steady decline in site temperatures and a cessation of height growth. In general, seedlings and emblings had similar ψ_{sat} and ψ_{tip} values through the growing season, and also had similar shifts in seasonal patterns of maximum modulus of elasticity, sympalstic fraction, and relative water content at turgor loss point.

Grossnickle and Major (1992) found that year-old and current-year needles of both seedlings and emblings had a similar decline in needle conductance with increasing vapour pressure deficit. Response surface models of current-year needles net photosynthesis (P_n) response to vapour pressure deficit (VPD) and photosynthetically active radiation (PAR) showed that emblings had 15% greater P_n at VPD of less than 3.0 kPa and PAR greater than 1000 µmol $m^{-2}s^{-1}$. Year-old and current-year needles of seedlings and emblings showed similar patterns of water use efficiency.

Rates of shoot growth in seedlings and emblings through the growing season were also similar to one another. Seedlings had larger shoot systems both at the time of planting and at the end of the growing season. Seedlings also had greater root development than emblings through the growing season, but root:shoot ratios for the 2 stock types were similar at the end of the growing season, when the survival rates for seedlings and emblings were 96% and 99%, respectively.

Problems Associated with Somatic Embryogenesis

- High chance of mutations.
- Difficult method.
- Loss of regenerative ability.
- High percentage of albino shoots during regeneration.
- Not possible with all plant species and must be optimized for each species and its use.

Tracking and Fate Maps

Understanding the formation of a somatic embryo through establishment of morphological and molecular markers is important for construction of a fate map. The fate map is the foundation in which to build further research and experimentation. Two methods exist to construct a fate map: synchronous cell-division and time-lapse tracking. The latter typically works more consistently because of cell-cycle-altering chemicals and centrifuging involved in synchronous cell-division.

Angiosperms

Embryo development in angiosperms is divided into several steps. The zygote is divided asymmetrically forming a small apical cell and large basal cell. The organizational pattern is formed in the globular stage and the embryo then transitions to the cotyledonary stage. Embryo development differs in monocots and dicots. Dicots pass through the globular, heart-shaped, and torpedo stages while monocots pass through globular, scuetellar, and coleoptilar stages.

Many culture systems induce and maintain somatic embryogenesis by continuous exposure to 2,4-dichlorophenoxyacetic acid. Abscisic acid has been reported to induce somatic embryogenesis in seedlings. After callus formation, culturing on a low auxin or hormone free media will promote somatic embryo growth and root formation. In monocots, embryogenic capability is usually restricted to tissues with embryogenic or meristematic origin. Somatic cells of monocots differentiate quickly and then lose mitotic and morphogenic capability. Differences of auxin sensitivity in embryogenic callus growth between different genotypes of the same species show how variable auxin responses can be.

Carrot *Daucus carota* was the first and most understood species with regard to developmental pathways and molecular mechanisms. Time-lapse tracking by Toonen et al.

(1994) showed that morphology of competent cells can vary based on shape and cytoplasm density. Five types of cells were identified from embryonic suspension: spherical cytoplasm-rich, spherical vacuolated, oval vacuolated, elongated vacuolated, and irregular shaped cells. Each type of cell multiplied with certain geometric symmetry. They developed into symmetrical, asymmetrical, and aberrantly-shaped cell clusters that eventually formed embryos at different frequencies. This indicates that organized growth polarity do not always exist in somatic embryogenesis.

Gymnosperms

Embryo development in gymnosperms occurs in three phases. Proembryogeny includes all stages prior to suspensor elongation. Early embryogeny includes all stages after suspensor elongation but before root meristem development. Late embryogeny includes development of root and shoot meristems. Time-lapse tracking in Norway Spruce *Picea abies* revealed that neither single cytoplasmic-rich cells nor vacuolated cells developed into embryos. Proembryogenic masses (PEMs), an intermediate between unorganized cells and an embryo composed of cytoplasmic-rich cells next to a vacuolated cell, are stimulated with auxin and cytokinin. Gradual removal of auxin and cytokinin and introduction of abscisic acid (ABA) will allow an embryo to form. Using somatic embryogenesis has been considered for mass production of vegetatively propagated pine clones and cryopreservation of germplasm. However, the use of this technology for reforestation and breeding of pine trees is in its infancy.

In Vitro Somatic Embryogenesis

In vitro somatic embryogenesis (SE) was first demonstrated in 1958 by Reinert and Steward. There are two ways by which SE could be obtained - i) Indirect SE, where first the callusing is induced from the explant by rapid cell division and later the callus give rise to SE, and ii) Direct SE, where the somatic embryos are developed directly from the explant without an intermediate callus phase.

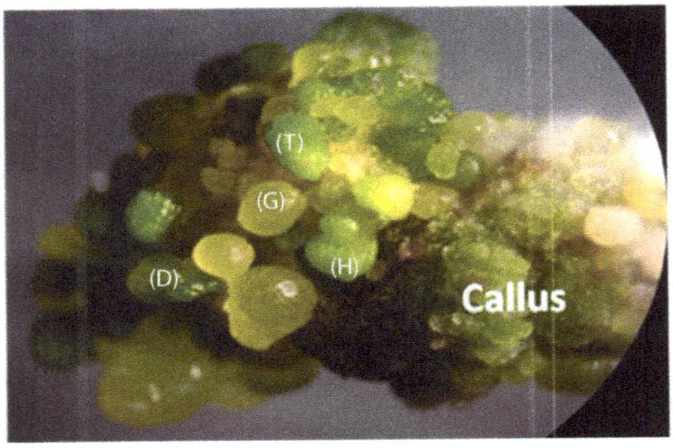

Somatic embryogenesis via callusing showing the development of globular(G), heart (H), torpedo (T) and dicot embryos (D) (arrow marked).

Direct somatic embryogenesis from cotyledon explant showing embryos at various stages of development.

In either of the cases, the somatic embryos resemble the zygotic embryos. In dicotyledonous plants, the somatic embryos passes through the globular, heart, torpedo and cotyledonary stages, as happens in zygotic embryos. The embryos germinate and develop into complete plantlets. The only major difference between somatic and zygotic embryogenesis is that somatic embryos do not pass through the desiccation and dormancy phases as happens in zygotic embryos, but rather continue to participate in the germination process.

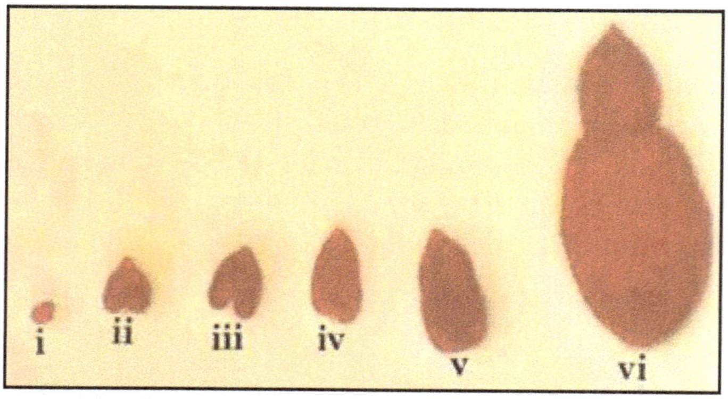

Different stages of development of zygotic embryos: (i) globular, (ii) early heart shape, (iii) late heart shape, (iv) torpedo shape, (v) early dicot, and (vi) fully developed dicot embryo.

Whether originating directly or indirectly via callusing, somatic embryos arise from single special cells located either within clusters of meristematic cells in callus mass or in the explant tissue. Somatic embryogenesis is regarded as a three step process:

i. Induction of embryo.

ii. Embryo development.

iii. Embryo maturation.

Organogenesis Versus Embryogenesis

In tissue cultures, plant regeneration via somatic embryogenesis may offer many advantages over organogenesis, such as:

i. Embryo is a bipolar structure rather than a monopolar one.

ii. The embryo arises from a single cell and has no vascular connection with maternal callus tissue or the cultured explant. On the other hand during organogenesis shoots or roots develop from a group of cells resulting into chimera formation which later establish a strong connection with the maternal tissue.

iii. Further, induction of somatic embryogenesis requires a single hormonal signal to induce a bipolar structure capable of forming a complete plant, while in organogenesis, it requires two different hormonal signals to induce shoot first and then root organ.

Factors Affecting Somatic Embryogenesis

Genotype and Type of Explant

Like organogenesis, SE is also genotype dependent for a given species and significant variations in response between cultivars have been observed in several plants like, wheat, barley, soyabean, rice, alfalfa etc. Genotypic variations could be due to endogenous levels of hormones, therefore, if the species has not shown SE previously, then it is required to test number of different cultivars of that species.

The next problem comes up is what tissue should be used as an explant in a particular species to induce SE? It can be decided very easily by closely examining of what explants were used in related species, genus or family. The explant selection is much more important than the media selection for SE process. Immature zygotic embryos have proved to be the best explant to raise embryogenic cultures as somatic embryos will form more readily from cells that are already in embryonic state. In *Azadirachta indica* (neem), the immature zygotic embryo at different stages of development, viz. globular, early to late heart shape, torpedo shape and early dicotyledonous stage, when cultured showed varied potential for SE. The globular embryos did not show any response. The older embryos germinated, formed calli or differentiated three types of organized structures, viz. shoots, somatic embryos and neomorphs (abnormal or embryo-like structures with varied morphology). Often the same explant differentiated more than one kind of regenerants. The most responsive stage of embryos was early dicotyledonous, followed by torpedo shaped embryos.

Growth Regulators

Auxin : Auxin plays a major role in the development of somatic embryos. All the well-studied somatic embryogenic systems, such as carrot, coffee and most of the cereals require a synthetic auxin for the induction of SE followed by transfer to an auxin-free medium for embryo differentiation. The synthetic auxin 2,4-D is the most commonly used auxin for the induction of SE. Besides, other auxins, NAA, IBA, picloram (4-Amino-3,5,6-trichloro-2-pyridinecarboxylic acid) and IAA, have also been used. A naturally occurring auxin IAA is a weak auxin and more readily broken down compare to 2,4-D and NAA. The auxins, particularly 2,4-D, in the concentration range of 0.5 – 1.0 mgl^{-1} (proliferation or induction medium), stimulates the formation of localized group of meristematic cells in the callus called 'proembryogenic masses' (PEMs), which are cell clusters within cell population competent to form somatic embryos.

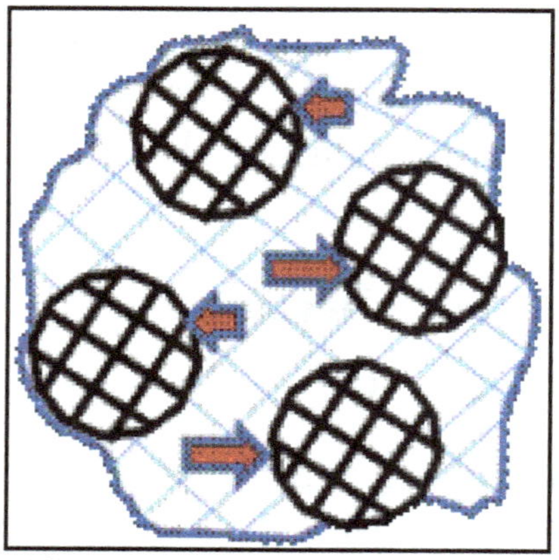

Embryogenic callus with PEMs (indicated by arrows) in the induction medium.

In repeated subcultures on the proliferation medium, the embryogenic cells continue to multiply without the appearance of embryos. However, if the PEMs are transferred to a medium with a very low level of auxin (0.01-0.1 mgl^{-1}) or no auxin in the medium (embryo development medium ; ED medium), they develop into embryos. The presence of an auxin in the proliferation medium seems essential for the tissue to develop embryos in the ED medium. The tissues maintained continuously in auxin-free medium would not form embryos. Therefore, the proliferation medium is called the 'induction medium' for SE and each PEMs as an unorganized embryo.

Cytokinin : There are reports of somatic embryo induction and development in cytokinin containing medium, but these reports are very few compared to those reporting induction by auxin alone or auxin plus cytokinin. Cytokinin, in general, induced SE directly without the callusing of explant. In most cases, TDZ is used as cytokinin, a herbicide, which mimics both auxin and cytokinin effects on growth and differentiation. The other cytokinins are also used when zygotic embryos are used as the explant source. The most commonly used cytokinins are BAP and Zeatin.

In *Azadirachta indica,* somatic embryo differentiation was influenced by the culture medium as well as the stage of embryo at culture. Maximum somatic embryogenesis occurred directly from the explant on BAP containing medium when early dicotyledonous stage of embryos were cultured. Medium with 2,4-D induced only neomorph differentiation directly from the explant. While torpedo shaped embryos showed both neomorph formation as well as somatic embryogenesis on BAP containing medium.

An explant showing differentiation of neomorphs (NEO) and somatic embryos (SE) on the same explant.

Neomorphs were suppressed embryos with green, smooth, shiny surface and solid interior. Although they were epidermal in origin like somatic embryos with heart shape notch but showed monopolar germination and no clear cut radicular region.

A. An explant showing direct differentiation of neomorphs. Some of these structures also show cotyledon-like flaps. The portion of the explant in contact with the medium has proliferated into a brownish green callus.

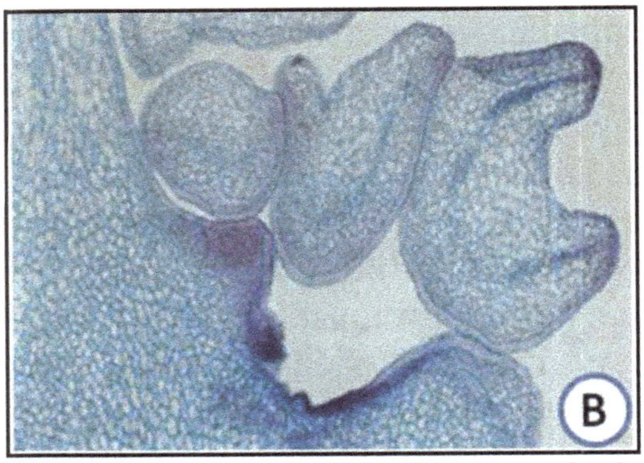

B. A histological section of A, showing epidermal origin of a neomorph of various shapes. It has a well differentiated epidermis and compactly arranged internal cells. These structures are loosely attached to the explant and show provascular strands.

Nitrogen Source

The most important nutrient of the culture medium is nitrogen which affects SE significantly. The form of nitrogen have a strong influence on the induction of SE. Often the presence of ammonium or some other source of reduced nitrogen is required, such as glycine, glutamate or casein hydrolysate. The ratio of ammonium to nitrate has also been shown to affect SE. In few cases, the calli initiated on a medium with KNO_3 as the sole source of nitrogen failed to form embryos upon removal of auxin. However, the addition of a small amount (5mM) of nitrogen in the form of NH_4Cl in the presence of 55mM KNO_3 allowed embryo development.

Embryo Maturation and Germination

Germination of somatic embryos can occur only when it is mature enough to have functional shoot and root apices capable of meristematic growth. Somatic embryos show poor germinable quality with respect to their convertibility into complete plantlets. This is because these embryos do not go through 'embryo maturation' phase which is the characteristic of seed or zygotic embryos. During this phase, accumulation of embryo-specific reserve food materials and proteins imparts desiccation tolerance to seed embryos and thereby promote their normal development for germination. Abscisic acid (ABA), which prevents precocious germination and promotes normal development of embryos by suppression of secondary embryogenesis and pluricotyledony, is reported to promote embryo maturation in several species. A number of other factors, such as temperature, shock, osmotic stress, nutrient deprivation and high density inoculums, can substitute for ABA, presumably by inducing the embryos to synthesize the hormone. ABA is known to trigger the expression of genes which normally express during the drying down phase of seeds. Probably the products of these genes impart desiccation tolerance to the embryos.

Secondary Somatic Embryogenesis

Secondary SE is a process in which new somatic embryos are proliferated from originally formed primary somatic embryos. Secondary SE is reported to have some advantage over primary somatic embryogenesis, such as high multiplication rate, long term repeatability and independency of an explant source. By repeated secondary SE selected embryogenic lines can be maintained for long period, in large quantities until the lines have been tested in field conditions particularly in perennial plants. Secondary SE also overcomes post fertilization barriers of the embryo, immature embryos of interspecific plants from incompatible crosses may be rescued by culturing them for secondary SE. It can also be used for the production of somatic embryos of species in which the embryos are the reservoir of important secondary metabolites. In *Azadirachta indica*, for secondary SE, primary embryos were used as the explant and when cultured on medium with TDZ and GA_3, secondary embryos were differentiated directly from hypocotyls region without any intervening callus. Whereas, a combination of BAP and IAA resulted into secondary SE preceded by callusing of the primary somatic embryo.

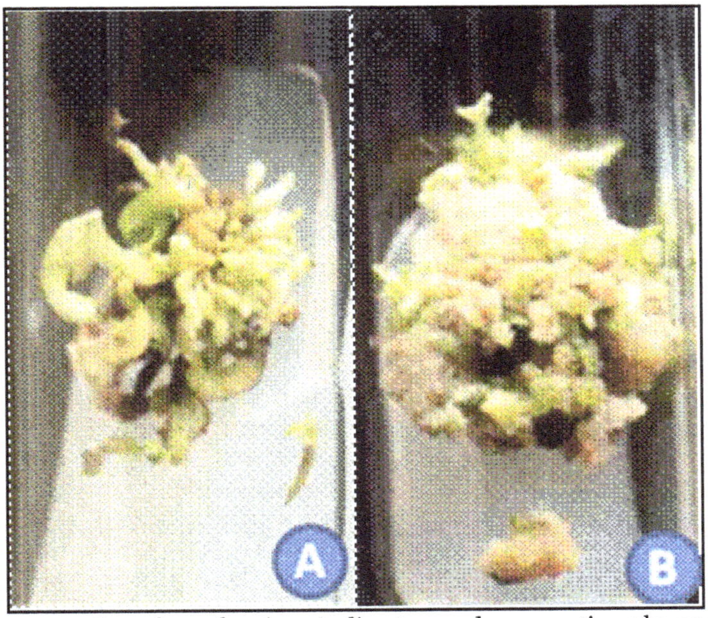

Primary somatic embryo showing: **A**. direct secondary somatic embryogenesis **B**. indirect secondary somatic embryogenesis.

Synchronization of Embryo Development

Generally, the differentiation of somatic embryos in semi-solid medium or liquid medium is highly asynchronous which adversely affect the germination rate. Synchronization of embryo development is very important for artificial seed technology. Of the several approached tried to achieve this, the most effective method are the physical separation of embryogenic stages and use of growth regulators to physiologically synchronize the development. The other alternative methods are the fractionation of embryos of different stages by filtration of suspension through meshes of different sizes or by gradient centrifugation. Besides, the most effective method to achieve synchronous development of somatic embryos is the use of substances that would induce reversible cessation of embryo development at a particular stage. ABA at low concentration is the most satisfactory substance for the purpose. For example, in carrot it inhibits the growth of roots and enhances suspension with torpedo shaped embryos.

Production of Synthetic Seeds or Artificial Seed

Although it is possible to use naked embryos for large scale planting, it would be beneficial to convert them into 'synthetic seeds' or 'synseeds' for large scale clonal propagation at commercial level. This can be achieved by encapsulating the viable somatic embryos in a protective covering. The coating material should have several qualities:

i. It must be non-damaging to the embryos.

ii. The coating should be mild enough to protect the embryos and allow germination but it must be sufficiently durable for rough handling during manufacture, storage, transportation and planting.

iii. The coat must contain nutrients, growth regulators and other components necessary for germination.

iv. The quality of somatic embryo should be good enough, they all are of uniform stage with reversible arrested growth and with high rate of conversion to plantlets.

Two types of synthetic seeds are produced:

I. Desiccated synthetic seeds II. Hydrated synthetic seeds.

I. Desiccated synthetic seeds : It involves encapsulation of somatic embryos followed by their desiccation and can be prepared by following methodology:

Mix equal volumes of embryo suspension + Polyox (polyoxyethylene)

↓

Suspension was dispensed on to a Teflon sheets (dried suspension sticks to glass plate)

↓

Dried to wafers in a laminar flowhood for about 5h until the wafers get seprated from teflon plate

The polyox is readily soluble in water and dries to thin film. It does not support the growth of microorganism and is non toxic to the embryos. Embryo survival and conversion of seeds are determined by redissolving the wafers in embryogenic medium and culturing the rehydrated embryos.

II. Hydrated synthetic seed: Several methods have been examined to produce hydrated artificial seeds of which Ca-alginate encapsulation has been the most widely used. It can be prepared by following steps:

Mix somantic embryos with Na-alginate

↓

Drop the mixture, using a pipette, into a 100mM solution of calcium nitrate

↓

Ion exchange reaction occurs and sodium ions are replaced by calcium ions forming Ca-alginate beads

Applications of Somatic Embryogenesis

Following features of somatic embryos prompted many scientists to achieve regeneration via somatic embryogenesis using various explants, most popular ones being zygotic embryos, or excised cotyledons or hypocotyls:

i. Somatic embryogenesis offers immense potential to speed up the clonal propagation of plants being bipolar in nature.

ii. Being single cell in origin, there is a possibility to automate large scale production of embryos in bioreactors and their field planting as synthetic seeds.

iii. The bipolar nature of embryos allows their direct development into complete plantlet without the need of a rooting stage as required for plant regeneration via organogenesis.

iv. Epidermal single cell origins of embryos favor the use of this process for plant transformation.

v. It can also be used for the production of metabolites in species where embryos are the reservoir of important biochemical compounds.

vi. The production of artificial seeds using somatic embryos is an obvious choice for efficient transport and storage.

vii. The embryo culture technique is applied to overcome embryo abortion, seed dormancy and self-sterility in plants.

Limitations of Somatic Embryogenesis

i. Complete conversion into plantlets or poor germination of embryos is a major limitation of somatic embryogenesis in many plants. Therefore, the process of germination needs to be studied in detail for successful plantlet conversion.

ii. Compared to other plant species active research on somatic embryogenesis involving forest trees has been very slow.

iii. The paucity of knowledge controlling somatic embryogenesis, the synchrony of somatic embryo development and low frequency of true to type embryonic efficiency are responsible for its reduced commercial application.

iv. To obtain a complete conversion into plantlets it is necessary to provide optimum nutritive and environmental conditions.

Androgenesis

In androgenesis, the male gametophyte (microspore or immature pollen) produces haploid plants. The basic principle is to stop the development of pollen into a gamete (sex cell) and force it to develop into a haploid plant or sporophyte. The remarkable discovery that haploid embryos and plants can be produced by *in vitro* culture of anthers of *Datura* (Guha and Maheshwari 1964,1966) brought renewed interest to haploidy. This method of androgenic haploid production was quickly attempted in many species to hasten the breeding programme in several economically important plants. Haploid production through anther/microspore culture scores higher over other methods due to the fact that anthers harbour large numbers of haploid microspores per anther and is a potentially efficient means to generate homozygous true-breeding progeny lines in plant breeding programs.

Methodologies

In androgenesis, immature pollen grains are induced to follow the sporophytic mode of development by the application of various physical and chemical stimuli. There are two methods for *in vitro* production of androgenic haploids –

(A) Anther culture, and (B) Isolated pollen (microspore) culture.

A. Anther culture:

(a) Select the flower buds from an elite plant and determine the stage of microspores by acetocarmine squashes of anthers or by staining them with fluorescent dye DAPI (4,6-diamidino-2-phenylindole).

(b) Surface sterilize the selected size of buds (ca 2mm size flower bud) to initiate in vitro anther cultures.

(c) Dissect the buds under a stereo-microscope, using pre-sterilized Petriplates, forceps and fine needles. Discard the damaged anthers, if any, and remove the filament gently.

(d) Inoculate the anthers, bearing early-to-late uninucleate stage of microspores, in a nutrient medium and maintain the cultures in defined conditions.

(e) As the anthers proliferate, they produce embryos/callus.

(f) The callus/embryos formed can be transferred to a suitable medium to finally produce haploid plants and then diploidize them by colchicine to produce homozygous diploids.

B. Microspore (Pollen) culture:

Haploid plants can also be produced from isolated immature pollens or microspores (male gametophytic cells):

(a) Select the flower buds from an elite plant and determine the stage of microspores by acetocarmine squashes of anthers or by staining them with fluorescent dye DAPI (4,6-diamidino-2-phenylindole).

(b) Surface sterilize the selected size of buds to initiate *in vitro* anther cultures.

(c) Extract the microspores by pressing and squeezing the buds with a glass rod against the sides of a beaker.

(d) Filter the pollen suspension to remove anther tissue debris.

(e) Wash and collect the viable and large pollen (smaller pollen do not regenerate) by filtration.

(f) Culture these microspores on a solid or liquid medium.

(g) As the microspores undergo multiple divisions, they produce multicellular and multinuclear structure.

(h) The callus/ embryos formed can be transferred to a suitable medium to finally produce a haploid plants and then diploid plants by colchicine treatment.

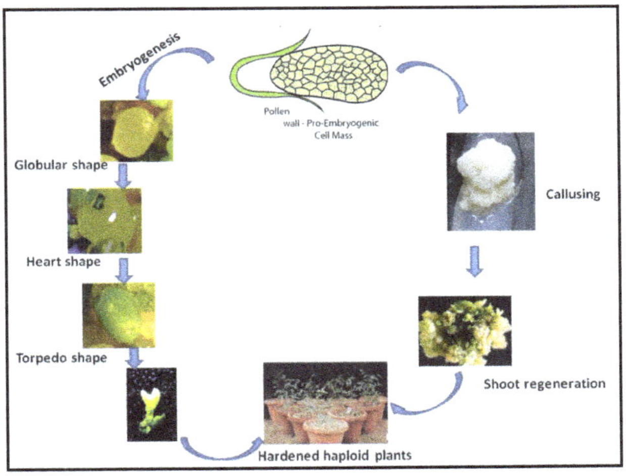

Androgenic haploid production pathways.

Comparison between Anther and Pollen Culture

Anther culture is an easy, quick and practicable approach. Anther walls act as conditioning factors and promote culture growth. Thus, anther cultures are reasonably efficient for haploid production. The major limitation is that the plants not only originate from pollen but also from other somatic parts of the anther. This results in the development of plants at different ploidy levels viz., haploids, diploids, aneuploids, as a result of which the final tissue, derived, may not be of purely gametophytic origin. Moreover, the plants arising from an anther would constitute a heterogenous population. It has been observed in some species that anther cultures show asynchronous pollen development, the older grains may suppress the androgenic potential of younger grains by releasing toxic substances . The disadvantages associated with anther culture can be overcome by pollen culture as it offers the following advantages:

- Undesirable effects of anther wall and associated tissues can be avoided.
- Androgenesis, starting from a single cell, can be better regulated.
- Isolated microspores (pollen) are ideal for various genetic manipulations like transformation, mutagenesis etc.
- The yield of haploid plants is relatively higher.

Pathways of Development

The early divisions in responding pollen grains may occur in any one of the following four pathways:

i) Pathway I - The uninucleate pollen grain may divide symmetrically to yield two equal daughter cells both of which undergo further divisions e.g. *Datura innoxia.*

ii) Pathway II - In some other cases e.g. *N.tabacum*, barley, wheat etc., the uninucleate pollen divides unequally. The generative cell degenerates, callus/embryo originates due to successive divisions of the vegetative cell.

iii) Pathway III - But in few species, the pollen embryos originate from the generative cell alone; the vegetative cell either does not divide or divides only to a limited extent forming a suspensor like structure.

iv) Pathway IV - Finally in few other species e.g. *Datura innoxia*, the uninucleate pollen grains divide unequally, producing generative and vegetative cells, but both these cells divide repeatedly to contribute to the developing embryo/callus.

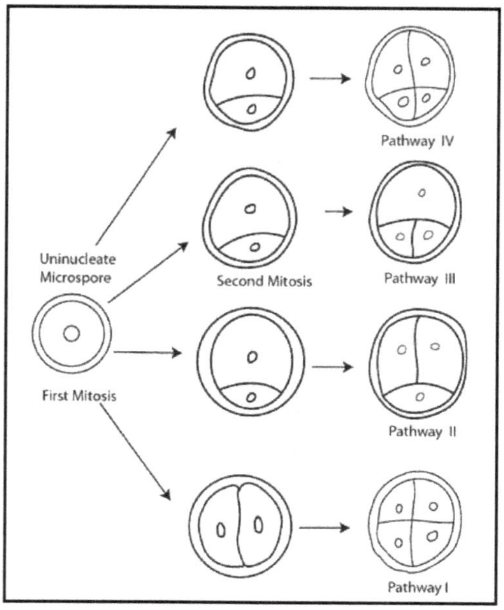

Pathways of development of microspores.

Gynogenesis

Gynogenic development of plants from unfertilized cells of female gametophyte (embryo-sac) in ovary/ovule/young flower cultures is one of the available alternatives for haploid production. It was first reported in barley San Noeum (1976). This method of haploid production is more tedious than androgenesis. The reasons for this being the indefinite numbers of microspores (male gametes) within the anther wall for androgenesis as against single egg cell (female gamete) per flower for gynogenic haploid production, which too, is deep seated within the embryo-sac (female gametophyte), thus making the entire process very cumbersome. The technique is very useful where anther culture has been unsuccessful, plants are male sterile or androgenesis is confronted with the problem of albino or non-haploid formation. The following techniques are generally used for production of haploids via *in vitro* gynogenesis either through direct embryogenesis or via callusing.

In Situ Parthenogenesis Induced by Irradiated Pollen Followed by in Vitro Embryo Culture

Parthenogenesis induced *in vivo* by irradiated pollen, followed by *in vitro* culture of embryos can be an alternative method of obtaining haploids in fruit crops. Gynogenesis by *in situ* pollination with irradiated pollen has been successfully used for *Malus domestica* (L.) Borkh, *Pyrus communis* L., *Actinidia deliciosa* (A. Chev). This method is based on *in vitro* culture of immature seeds or embryos, obtained as a result of pollination by irradiated pollen with gamma rays from cobalt 60. The method is useful in those species in which *in vitro* anther culture has not been successful. Irradiation does not hinder pollen germination but prevents pollen fertilization and, thereby, stimulating the development of haploid embryos from ovules. The success of this technique is dependent on the choice of radiation dose, the developmental stage of the embryos at the time of culture, culture conditions and media requirements.

Ovary Slice Culture

Ovary slice culture technique involves culture of transverse sections of unpollinated ovaries on culture media. The following protocol was used to induce *in vitro* gynogenesis in Tea (Hazarika and Chaturvedi 2012):

a. For Ovary slice culture in Tea, unopened and unpollinated mature flower-buds (6-10 mm) size were collected early in the morning. Some of the buds were fixed in FAA (5:5:90 v/v/v Formaldehyde: Acetic acid: 70% Ethanol), for 48 h, and then stored in 70% alcohol. Later on, the appropriate developmental stage of the embryo sac was determined by histological analysis.

b. The flower buds were surface sterilized with 0.1% $HgCl_2$ for 7 minutes, followed by rinsing with sterile distilled water at least thrice.

c. Carefully dissected transverse sections of ovaries were cultured on Murashige and Skoog's media supplemented with varying concentrations of Auxins and Cytokinins.

d. Six ovary slices containing unpollinated ovules were cultured in 60 mm X 15 mm pre sterilized disposable Petriplates containing 10 ml MS medium.

e. The sealed Petriplates were subjected to various regimes of temperature and light treatments.

Ovule Culture

The unfertilized ovary is surface sterilized and the ovules were taken and placed into culture. Excision of ovule, followed by culture on specific media may be either extremely easy to accomplish, as in case of large-seeded species in which only a single ovule is present, or time-consuming and intricate, in small-seeded polyovulate species. Two types of ovule support systems have been developed. The filter paper support system involves culturing of the ovules on top of filter paper placed over liquid medium, whereas the vermiculite support technique demands placing the ovules on a sterile vermiculite/liquid media mixture (vermiculite support) with the micropylar side down. Unpollinated ovule culture has been used for haploid production in sugar beets and onions.

Factors Affecting Haploid Androgenesis

There are numerous endogeneous and exogeneous factors that affect *in vitro* haploid production. These factors can be genetic, physiological, physical and chemical may also interact amongst each other to divert the microspores/egg cell to enter into a new developmental pathway. Some of the crucial factors affecting haploid production in plants have been discussed below:

Genotype of the Donor Plant

The genotype of the donor plants, i.e. genetic factor, has a great influence on the anther, ovary and ovule culture response. In earlier studies, significant difference in callus formation using varieties or crosses were observed. In some species only a few genotypes have responded of many tested. In fact genetic factors contribute in a major way to the differences in the number of haploid plants produced (Custódio et al, 2005).

Physiological Status of the Donor Plant

The physiological conditions of the donor plant, *i.e* the environmental conditions and age of the donor plant, directly affects both *in vitro* androgenesis and *in vitro* gynogenesis in almost all plant species. A correlation between plant age and anther response has also been demonstrated. Similar is the case with ovary culture. The frequency of androgenesis is usually higher in anthers harvested at the beginning of the flowering period and showed a gradual decline in relation to plant age (Bhojwani and Razdan 1996). Varying temperature and light conditions during the growth of donor plants also affect anther response. In anther culture of grape, the induction frequency of embryoids derived from spring flowers was higher than that derived from summer flowers (Zhou and Li 1981). The microscopical observations showed that some varieties of rubber tree often have a lot of degenerated and sterile microspores in their anthers in early spring or hot summer due to the influence of unfavourable climatic conditions (Chen et al, 1982). As a result no pollen embryoids were obtained from such anthers but only the somatic calli.

Stage of Explants Material at the Time of Inoculation

Stage of Microspores

The stage of microspores at the time of inoculation is one of the most critical factors in the induction of androgenesis. Detailed cytological studies conducted on poplars, rubber (Chen 1986) and apple (Zhang et al, 1990) have shown that androgenic callus and embryos were mainly induced through a deviation of the first pollen mitosis to produce two undifferentiated nuclei. Besides affecting the overall response, the microspore stage at culture also has a direct bearing on the nature of plants produced in anther culture (Sunderland and Dunwell 1977). About 80% of the embryos obtained from binucleate microspores of *Datura innoxia*, a highly androgenic species, were non-haploids (Sunderland et al, 1974). In a vast majority of species where success has been achieved, anthers were cultured when microspores were at the uninucleate stage of microsporogenesis (Chaturvedi et al, 2003; Pedroso and Pais 1994; Sopory and Munshi 1996).

Stage of the Embryo Sac

It has been reported that the effect of ovule development on gynogenesis is profound as it harbours the embryo sac comprising of the egg cell. The stage of embryo sac is an important determining factor for *in vitro* gynogenesis in various plant species. However, it is difficult to know the stage of embryo sac at the time of inoculation. Several authors prefer to describe the inoculation stage according to the developmental stage of the flower bud or stage of pollen development. However, this could not be possible in several species, where male and female gametophytes do not mature simultaneously, a phenomenon known as protandry, the maturation of anthers before carpels (e.g., onion, leek, sunflower, sugar beet and carrot) and the opposite protogyny (e.g., pearl millet). In such cases, the stage of embryo-sac at culture can be determined by histological preparations of ovary/ovules that are at identical stage with that of cultured ovary/ovules.

Although a wide range of embryo sac stages are responsive to gynogenic development, but, in most cases nearly mature embryo-sac stage gave better results. This is quite contrary to anther culture in which mature pollen is non responsive to androgenesis.

In Barley and rice, unfertilized ovary cultures with late staged mature embryo sacs gave good results (San Noeum 1976, 1979; Wang and Kuang 1981) while others reported success with ovary cultures containing uninucleate to mature embryo sacs (Zhou and Yang, 1981b, 1982; Kuo, 1982; Huang et al., 1982).

Anther Wall Factor

One of the important research subject in anther culture of woody plants is to avoid the over proliferation of callus from anther wall tissues and to achieve a high yield of pollen embryoids and pollen calli. In anther culture of most woody plants, both pollen calli (or embryoids) and somatic calli from anther wall tissues grew simultaneously. The development of callus from somatic tissues of anther can be avoided by culture of isolated microspores. However, microspore culture is not very successful in woody plants (Chaturvedi et al, 2003). Pelletier and Ilami (1972) had introduced the concept of "Wall Factor", according to which the somatic tissues of anther play an important role in the induction of sporophytic divisions in pollen. In anther culture of rubber, 47% of the microspores in close contact with the surrounding somatic cells could develop into multicellular masses as compared to only 5% of microspores away from the wall (Chen 1986). Anther wall callusing was regarded as a pre-requisite for the formation of androgenic haploids (Chaturvedi et al, 2003; Chen 1986; Chen et al, 1982).

Chemical Factors

The constituents of the basal medium and combination/s of growth regulators serve as important factors in eliciting successful androgenesis or gynogenesis. However, it is difficult to suggest one culture medium or one growth regulator for all the systems. The requirement of growth regulators and culture medium in terms of kind and concentration may vary with the system. Generally, there is an agreement that the source and amount of total nitrogen as well as a combination of a cytokinin and an auxin is necessary for pollen embryogenesis and pollen callusing in several woody plants (Chaturvedi et al, 2003; Chen 1986; Nair et al, 1983). Sucrose has generally been used as the major carbohydrate source in the culture medium. The effect of sucrose on anther culture has been investigated in a number of species. Generally, sucrose is supplied at 2-3% concentration. However, increase in its concentration can lead to beneficial morphogenic potential by suppressing the proliferation of somatic tissues.

Effect of Female Flower Position

Position of female flowers on the plant stem affected induction of embryos from ovule cultures of *Cucurbita pepo* L (Shalaby, 2007). One of the possible explanations for enhancing responses of tissue culture could be attributed to indigenous hormonal level (Johansson, 1986).

Diploidization of Haploids

The haploids may grow normally up to flowering stage but in the absence of homologous chromosomes the meiosis is abnormal and consequently, viable gametes are not formed. To obtain fertile homozygous diploids, the chromosome complement of the haploids must be duplicated. For long colchicine has been used for the purpose and is more effective.

Another method to diploidise the haploids is by utilizing the tendency of *in vitro* growing callus cells to undergo endomitosis. Segments from vegetative parts of haploid plants are grown on a suitable culture medium and made to proliferate into calli. After some time many of the callus cells become diploid due to endomitosis. By transferring such calli to an appropriate medium fertile diploids shoot can be obtained.

Applications of Haploid Production

Development of Pure Homozygous Lines

Homozygous, true breeding cultivars are highly important for screening of high yielding lines and to produce hybrid vigour as a method of crop improvement. Obtaining homozygous diploid plants by conventional methods is difficult in perennials. From several decades to over a hundred years are required to obtain a pure line by means of successive inbreeding throughout many generations. The seed set by inbreeding in many trees is very low, usually only a few of ten thousandth or sometimes no seed can be obtained at all; therefore, it is impractical to obtain pure lines by inbreeding (Chen, 1986). Moreover, conventional method of haploid production by inbreeding is impossible if the plant is strictly cross-pollinating in nature. On the other hand, homozygous diploid plants can be achieved in a single generation by diploidization of *in vitro* raised haploids by colchicine treatment.

Genetic Studies

Because of the lack of accurate materials in research work, the progress in the study of genetics in trees is much slower than that in annual herbaceous plants. The genetics of a lot of important traits in economically important plant species has not been clearly demonstrated as yet. As a result, it is still unknown whether the desirable characters of the parents will appear in their progenies. Only when crossing between different homozygous diploid plants is carried out, we can gain a clear idea of dominance of genes controlling various characters and that these characters are either monogenic or polygenic (Chen 1986). Furthermore, if we can use the haploid plants as samples of gametes, then we can obtain directly the recombination value between genes. Moreover, we can also use the haploid plants to study chromosome homology within genome or between genomes.

Gametoclonal Variation

The "gametoclonal variation" arises among plants regenerated from cultured gametic cells consisting of differences in morphological and biochemical characteristics as well as in chromosome number and structures that are observed. Besides yielding haploids, *in vitro* androgenesis helps in the screening of gametophytic variation at plant level. Pollen grains within an anther form a highly heterogeneous population because they are the product of meiosis which involves recombination and segregation. Therefore, each pollen plant is genetically different from the other. The gametoclonal variants being hemizygous in nature expresses also the recessive characters in the R_0 plants (Bhojwani and Razdan, 1996). Different sources of variation can explain gametoclonal variation such as new genetic variation induced by cell culture procedures, from segregation and independant assortment, chromosome doubling procedures etc (Morrison and Evans, 1987; Huang, 1996).

Induction of Mutations

In general, majority of induced mutations are recessive and, therefore, are not expressed in diploid cells due to the presence of dominant allele. Since, haploid plants have only one set of chromosomes, their dominant and recessive characters can be seen simultaneously on separate plants. It is extremely advantageous to provide a convenient system for the induction of mutations and selection of mutants with desirable traits in the absence of their dominant counterparts (Bhojwani and Razdan, 1996).

Obtaining New Genotypes with Alien Chromosomes

The technique of interspecific and intergeneric hybridization can be combined with anther culture techniques (Thomas et al., 2003) for obtaining new genotypes with alien chromosomes. Thus, new genotypes with various reconstructed chromosome complements can be obtained after their successful chromosome doubling.

Genetic Manipulation

As microspore culture is a single cell system, it makes selection at the single cell level possible and, furthermore, offers new prospects for genetic manipulation like mutagenesis and transformation. Direct gene transfer by microinjection offers the possibility of transgenic plant formation by using isolated pollen culture having high regeneration efficiency (Kasha and Maluszynski, 2003). Moreover, if transgenes can be incorporated into the haploid microspore genome, prior to DNA synthesis and chromosome doubling, the doubled haploids may also be homozygous for the transgenes. Thus, isolated microspores not only provide a good target for bombardment but, also are readily amenable to transgene *in vitro* selection. Jahne et al. (1994) were the first to achieve plants homozygous for the transgenes using biolistic bombardment of barley microspores.

Genomics

Doubled haploids play a vital role in genomics, especially, in the integration of genetic and physical maps, thereby, providing precision in targeting candidate genes (Kunzel et al., 2000; Wang et al., 2001). Doubled haploids combined with marker assisted selection provides a short cut in backcross conversion, a plant breeding method for improving an elite line defective in a particular trait (Toojinda et al., 1998). Expressed sequence tags may help in identification of genes that determine any trait of interest.

Triploid Production by Endosperm Culture

Introduction

Endosperm is a unique tissue in its origin, development and ploidy level. It is a product of double fertilization but unlike the embryo it is triploid and develops into a formless tissue (Bhojwani and Bhatnagar 1999). It is, therefore, an interesting tissue for morphogenesis. Any abnormality in the development of endosperm may cause the abortion of embryo resulting in sterile seeds (Johnston et al. 1980). The endosperm may be totally consumed by developing embryo leading to the formation of exalbuminous (non-endospermous) seeds or when it persists, the seed is called albuminous (endospermous). In albuminous seeds, it is used as a food source which may contain proteins, starch or fats and the embryo can utilize this food during seed germination.

Cellular totipotency of endosperm cells was first demonstrated by Johri and Bhojwani in 1965. To date, differentiation of shoots/embryos/plantlets from endosperm tissue has been reported for more than 64 species belonging to 24 families. In many of these reports the regenerants were shown to be triploid. A key factor in the induction of cell divisions in mature endosperm cultures is the association of embryo. The embryo factor is required only to trigger cell divisions; further growth occurs independent of the embryo. Triploid plants are usually seed-sterile. However, there are many examples where seedlessness caused by triploidy is of no serious concern or, at times, even advantageous. Some of the crops where triploids are already in commercial use include several varieties of apple, banana, mulberry, sugar beet and watermelon. Natural triploids of tomato produced larger and tastier fruits than their diploid counterparts (Kagan-Zur et al. 1990).

Traditionally, triploids are produced by crossing induced superior tetraploids and diploids. This approach is not only tedious and lengthy (especially for tree species) but in many cases it may not be possible due to high sterility of autotetraploids. The first step in the process is to produce tetraploids by colchicine treatment of germinating seeds, seedlings or vegetative buds. In most of these cases the rate of induction of tetraploids had been low (7-22%). Moreover, the treatment is lengthy and laborious. Once tetraploids have been produced, their cross with the diploid parent may not be successful in majority of the cases. In successful crosses the seed-set, seed germination and survival rate of the seedlings is low. Moreover, all sexually produced triploids do not behave uniformly, which may be due to segregation both at tetraploid level and subsequent population of crosses with putative diploid. In contrast, *in vitro* regeneration of plants from endosperm, a naturally occurring triploid tissue, offers a direct, single step approach to triploid production. The selected triploids, expected to be sexually sterile, can be bulked up by micropropagation.

Factors Controlling Callus Proliferation and Plant Regeneration

Endosperm at Culture

Usually culture of endosperm needs the selection of seeds at proper stage of development. This is usually calculated as days after pollination (DAP) and it varies from plant to plant as 9-10 days after pollination (DAP) in *Lolium perenne* (Norstog 1956), 8-11 DAP *in Zea mays* (Tamaoki and Ullstrup 1958), 8 DAP in *Triticum aestivum* and *Hordeum vulgare* (Sehgal 1974) and 4-7 DAP in *Oryza sativa* (Nakano et al. 1975). Usually free nuclear endosperm did not produce any callus and the intensity of response depends on the level of organization of endosperm cells.

Plant Growth Regulators and Other Supplements

Selection of a suitable basal medium and the addition of proper growth regulators and other adjuvants are the decisive factors that determine the success of triploid plant development. The culture of immature endosperm requires yeast extract (YE), casein hydrolysate (CH), coconut milk (CM), corn extract (CE), potato extract (PE), grape juice (GJ), cow's milk (CWM) or tomato juice (TJ) despite a suitable medium and growth regulators. Murashige and Skoog (1962) basal medium was mostly used to initiate and improve the response in *in vitro* endosperm cultures. White (1963) basal medium (WM) was also employed in some cases. La Rue (1949) used various organic supplements like CE, PE, TJ, GJ, YE or CWM to raise endosperm callus cultures. Of these, TJ was found to be superior over other additives due to cytokinin-like activities.

Later it was found that the TJ could be replaced by YE. YE induced callus proliferation was reported in *Zea mays, Croton, Jatropha panduraefolia, Lolium, Ricinus communis, Oryza sativa, Coffea arabica*, and *Juglans regia*. Other additives like CH in *Zea mays, Exocarpus cupressiformis, Dendrophthoe falcata, Nuytsia floribunda, Putrangiva roxburghii, Hordeum vulgare, Achras sapota, Citrus grandis, Prunus persica, Actinidia chinensis, Actinidia urguta, A. deliciosa* and CM in *Acacia nilotica*, and *Codiaeum variegatum* were also employed by different workers.

Most of the immature endosperm needs the presence of one or more growth regulators for plant regeneration except in few , where MS basal medium is sufficient for endosperm embryogenesis. In majority of reports, an auxin, preferably 2,4-D is necessary for callus induction from immature endosperms. In case of mature endosperm, the optimum callus growth was observed either on a cytokinin or a cytokinin in combination with an auxin and for autotrophic taxa, cytokinin, auxin, CH or YE is necessary. In most of the cases the time required to initiate proliferation varies from 10 days to 20 days, but pre-soaking of endosperms with GA_3 have reduced the time period from 10 days to 7 days.

Physical Factors

This includes effect of temperature, light and pH on endosperm proliferation. Straus and La Rue (1954) observed that corn endosperms develop better in dark than light conditions. But in *Ricinus* reverse is the case where a continuous light period was found optimum for endosperm proliferation (Srivastava 1971). In some cases, the endosperms were cultured along with the embryo and kept in the diffuse light with 16 h photoperiod. Light conditions facilitate the early germination of embryo and the embryos can be removed easily due to their characteristic green colour (Thomas et. al. 2000). In coffee, the endosperm callus grows better under 12 h light/dark conditions (Keller et al. 1972). In *Lolium* the light doesn't have any significant role on endosperm proliferation (Norstog 1956).

Not much research has been carried out till date with regard to the effect of temperature and pH on endosperm proliferation. In available literature the optimum temperature for endosperm growth was reported to be 25°C (Johri and Srivastava 1973). The pH varies from 4.0 for *Asimina* to 5.0 for *Ricinus* , 5.6 for *Jatropa* and *Putranjiva* and 6.1 for *Zea mays* . In general, 5.5 to 5.8 pH seem to support the best growth of endosperm tissues in cultures.

The Embryo Factor

There is an absolute necessity of the so called "embryo factor" for the proliferation of endosperm (Bhojwani 1968; Srivastava 1971a, b). Some factors contributed by the germinating embryo is required for the stimulation of mature and dried endosperms of few plant species. In general, it has been found that mature endosperm requires the initial association of embryo to form callus but immature endosperm proliferates independent of the embryo. However, in neem the association of the embryo proved essential to induce callusing of immature endosperm; the best explant was immature seeds (Chaturvedi et al. 2003). Similar observation for mulberry was reported by Thomas et al. (2000). However, the embryo factor can be overcome by the use of GA_3 as was observed in *Croton bonplandianum* and *Putranjiva roxburghii* (Srivastava 1973). It is reported that during germination, the embryo releases certain gibberellin like substances, which may promote the endosperm proliferation (Ogawa 1964; Ingale and Hageman 1965). However the mature endosperms of *Achras sapota, Santalum album, Emblica officinalis* and *Juglans regia* (Cheema and Mehra 1982; Tulecke et. al. 1988) could proliferate without the association of embryo or pre-soaking of them in GA_3.

Shoot Regeneration

Organogenesis from endosperm tissue was first reported in *Exocarpus cupressiformis* (a member of the family Santalaceae) by Johri and Bhojwani (1965). The pathway of plant regeneration includes shoot-bud differentiation or embryogenesis directly from the explants or indirectly from proliferating callus. In almost all the parasitic angiosperms, the endosperm shows a tendency of direct differentiation of organs without prior callusing, whereas in the autotrophic taxa the endosperm usually forms callus tissue followed by the differentiation of shoot buds, roots or embryos. Direct shoot regeneration from the cultured endosperm was observed in a number of semi-parasitic angiosperms including *Exocarpus*, *Taxillus*, *Leptomeria*, *Scurrula* and *Dendrophthoe* .

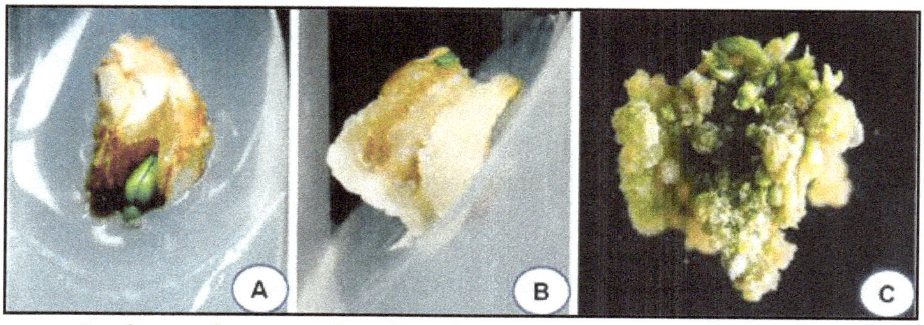

Shoot regeneration from endosperm callus of *Azadirachta indica* : **A.** An immature seed in culture has split open after 2 weeks and releasing the green embryo and callused endosperm , **B.** White fluffy endosperm callus can be seen from the fully opened seed after three weeks. Embryo is lying at one end of the explant, **C.** 6 -week-old subculture of endosperm callus showing the differentiation of distinct shoots and nodules as well.

In *Exocarpus*, an auxin (IAA) along with cytokinin (Kinetin) was required for direct shoot regeneration (Johri and Bhojwani 1965). Addition of zeatin in WM gave rise to green shoots from the intact seed (i.e. endosperm with embryo) culture of *Scurrula pulverulenta* which on subculture gave rise to characteristic haustoria (Bhojwani and Johri 1970). In *Taxillus vestitus*, shoot bud formation occurred on WM supplemented with IAA, Kinetin and CH, after seven weeks. Replacement of IAA with IBA could induce shoot regeneration in 55% cultures and haustoria in 60% cultures (Johri and Nag 1970). Here, the embryo had an adverse effect on bud differentiation from endosperm. Injury to the endosperm was found to be beneficial for shoot induction in some cases; shoot buds first develop along the injured region. The position of the explant on medium plays a significant role in regeneration of shoot in *Taxillus* spp. When half split *T. vestitus* endosperm without embryo was placed on medium with its cut surface in contact with the medium containing Kinetin, 100 % of the cultures produced shoots. In *Leptomeria acida*, IBA proved more efficient than IAA in terms of rapid callus proliferation. However, on IAA supplemented medium the callus gave rise to shoots in 100% cultures (Nag and Johri 1971).

The callus proliferation from endosperm and the subsequent shoot organogenesis was also reported in *Jatropa panduraefolia*, *Putranjiva roxburghii*, *Codiaeum variegatum*, *Malus pumila*, *Oryza sativa*, *Annona squamosa*, *Actinidia chinensis*, *Mallotus philippensis*, *Actinidia deliciosa*, *Morus alba*, *Azadirachta indica* and *Actinidia deliciosa*. In *Actinidia* species callus initiation occurred on MS medium supplemented with 2,4-D and Kinetin. Transfer of these calli to MS medium containing IAA and 2ip resulted in shoot and root organogenesis. In apple, endosperm proliferated into callus on MS medium supplemented with Kinetin + 2,4-D/BA + NAA and subsequent regeneration occurred on MS medium fortified with BA + CH. In *Annona squamosa* the callusing of endosperm occurred on WM supplemented with two cytokinins (Kinetin and BA), an auxin (NAA) and Gibberellic acid (GA_3). But organogenesis in the callus occurred on Nitsch's medium supplemented with BA and NAA.

In *Mallotus philippensis,* a continuously growing callus was obtained on MS medium supplemented with 2, 4-D + Kinetn. These calli when subcultured on MS + BA + CH gave rise to various morphologically distinct cell lines, of these, only the green compact cell line was responsive for organogenic differentiation. Shoot regeneration occurred in this callus when subcultured on MS medium fortified with BA + NAA.

In Rice, there was a striking difference in the growth response of immature and mature endosperm. Immature endosperm underwent two modes of differentition i.e. direct regeneration from the explant or indirectly via intervening callus phase. In mature endosperm, shoot organogenesis was always preceded by callusing. Callus from mature endosperm was initiated and maintained on MS + 2,4-D; shoot differentiation from callus occurred on MS + IAA + Kinetin. The proliferation of immature endosperm and occasional shoot formation occurred on YE supplemented medium; addition of IAA and Kinetin improved the response further.

Immature endosperms of neem (*Azadirachta indica*) showed best callusing on MS + NAA + BA + CH. When the callus was transferred to a medium containing BA or Kinetin, shoot buds differentiated from all over the callus. Maximum regeneration in terms of number of cultures showing shoot-buds and number of buds per callus cultures occurred in the presence of BA. Thomas et al. observed maximum callusing of Mulberry endosperm on MS + BA + NAA + CH or YE. Shoot buds were emerged when the callus was transferred on a medium containing a cytokinin or a cytokinin and a - naphthaleneacetic acid (NAA). The percent response was highest on BA and NAA containing medium. However, the number of shoots per explants was maximum when TDZ alone was used.

Histology

Histological studies of the proliferating endosperm of *Jatropa*, *Putranjiva* and *Ricinus*, revealed that the embryo also enlarged and proliferated along with the endosperm but soon showed the sign of degeneration. In such cases the endosperm calli were transferred to a fresh medium to avoid any contamination from degenerated embryonal cells. The 4-week-old callus derived from endosperm cultures, proliferated into parenchymatous cells and 6-week old callus showed tracheidal cells (Srivastava 1971a, b; Srivastava 1973;

Johri and Srivastava 1973). In *Santalum*, endosperm proliferation started after the formation of several meristematic layers below the epidermal region (Rangaswamy and Rao 1963). By carefully applying plant growth regulators the nodular outgrowths can be induced on the surface of the cultured endosperm as in case of *Osyris wightians* (Johri and Bhojwani 1965) and *Putranjiva roxburghii* (Srivastava 1973).

The importance of tracheidal differentiation in the callus of endosperm cultures is that it facilitates organogenic differentiation. In the families like Euphorbiaceae, Loranthaceae and Santalaceae, the endosperm tissues readily form tracheidal elements in cultures (Johri and Srivastava 1973, Johri and Bhojwani 1971). In *Emblica officinale*, tracheidal cells and cambium like cells organized into vascular strands or nodules in the differentiation medium while in the callusing medium tracheidal cells remained scattered. The differentiation of vascular strand in the callus accompanied the shoot bud formation (Sehgal and Khurana 1985).

In *Aleuritus fordii*, callus proliferated from endosperm explant consisted of large, compact and vacuolated cells. Tiny group of cells became distinct from adjoining large and vacuolated cells and became meristematic. These cells remained thin walled with dense cytoplasm and a large clear nucleus. Later the meristemoids developed in to dome shaped shoot apex, which produces leaf primordia (Syed Abbas 1993). In *Mallotus phillippensis* only the compact green callus underwent differentiation. Such callus showed vasculature, developed protuberances and eventually gave rise to shoots buds. Small group of cells with deep-seated distinct meristematic loci were also observed in these calli, which later gave rise to dome-shaped shoot primordia, endogenously (Sehgal and Abbas 1996).

In mulberry (*Morus alba*), histological analysis revealed that the older region of the callus comprised of highly vacuolated cells. Shoot buds differentiated from peripheral nodular structures, which comprised of compactly arranged highly cytoplasmic cells. Often a few layers of degenerating vacuolated cells were seen outside the shoot primordia. It is possible that the shoots originated from inside the nodules and emerged after rupturing the surrounding tissue. The regenerating shoots showed vascular supply continuous with the vasculature of the callus (Thomas 2000).

Both exogenous and endogenous differentiation of shoots was observed in *Azadirachta indica* . The serial section of two-week-old regenerating callus showed that many meristematic pockets developed from inside the callus, which developed into shoot buds after 3 weeks. Histological sections also revealed that the shoot buds emerged from the peripheral tissues of the callus as well (Chaturvedi et al. 2003). In *Actinidia deliciosa* histological analysis of the freshly isolated endosperm revealed small intercellular spaces and cells were filled with storage materials. However, the calli derived from the endosperm were larger, vacuolated and lacked storage materials. In older callus daughter nuclei attached to newly formed cell walls were often observed, suggesting disturbances of cell division. The cells differed in size and shape and contained nuclei with variable numbers of nucleoli (Goralski et al. 2005).

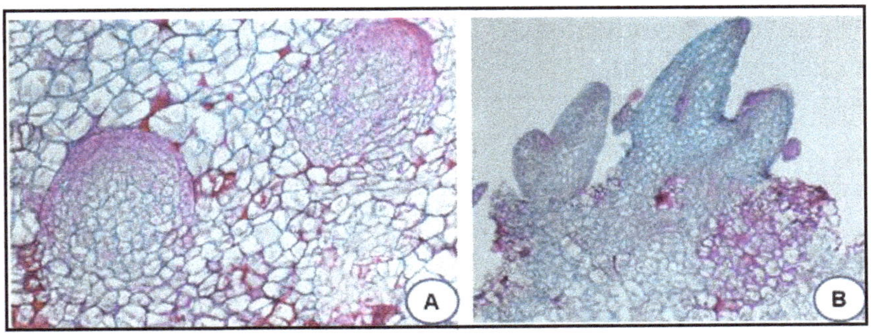

A. Section of a 2-week-old regenerating callus from endosperm of *A. indica* showing endogenous meristematic pockets. **B.** Section of a 4-week-old regenerating callus showing distinct shoot buds differentiated from peripheral vascularized nodule. One of the buds showing glands.

Cytology

The endosperm tissue often shows a high degree of chromosomal variation and polyploidy. Mitotic irregularities, chromosome bridges and laggards are other important characteristic features of endosperm tissue. Some reports suggest that the cells of endosperm cultures showed ploidy higher than 3n as in the case of *Croton* (Bhojwani and Johri 1971), *Jatropha* (Srivastava 1971a) and *Lolium* (Norstog et al. 1969). Cytological observations of the endosperm callus, derived from *Dendrophthoe falcata*, *Taxillus cuneatus* and *Taxillux vestitus*, showed diploid (2n=18) and triploid (3n=27) chromosomes (Johri and Nag 1974).

In addition to the cytological observations of endosperm callus, the chromosomal analysis of the regenerated plantlets were also studied in a number of systems. In *Juglans regia* two plants of endosperm origin were analysed for ploidy determination and both the plants showed triploid (3n=3x=48) number of chromosomes (Tulecke et al. 1988). In *Citrus* stability of the ploidy level and chromosome number were observed all through the regeneration process and triploid (2n=3x=27) plantlets were recovered (Gmitter et al. 1990). In *Mallotus philippensis* the squash preparation of root tip cells of the regenerated plants invariably showed triploid chromosome number (3n=3x=33) (Sehgal and Abbas 1996). The triploid nature of the endosperm-derived plants was determined by Feulgen cytophotometry in *Acacia nilotica* (Garg et al. 1996).

In Mulberry (*Morus alba*), 7-month-old plants of endosperm origin were utilized for ploidy determination. All the ten plants analysed cytologically showed triploid number of chromosome (2n=3x=42) (Thomas et al. 2000). The ploidy determination of 20 plants of *Azadirachta indica* , regenerated from endosperm calli, showed that 66% of the plants had triploid number of chromosomes (2n=3x=36) and the rest 34% were diploids (2n=2x=24) (Chaturvedi et al. 2003). In *Actinidia deliciosa* three different ploidy levels viz., 3C, 6C and 9C were observed in cells of endosperm derived callus analyzed by flow cytometry. The analysis of the leaves of endosperm derived plants showed 45.7% fluorescence intensity peaks corresponding exactly to 3C whereas 42.2% exhibited peaks of fluorescence intensity representing C-values between 2C and 4C.

Only 8.4% of the samples indicated 2C DNA content, and one sample showed 6C DNA content (Goralski et al. 2005).

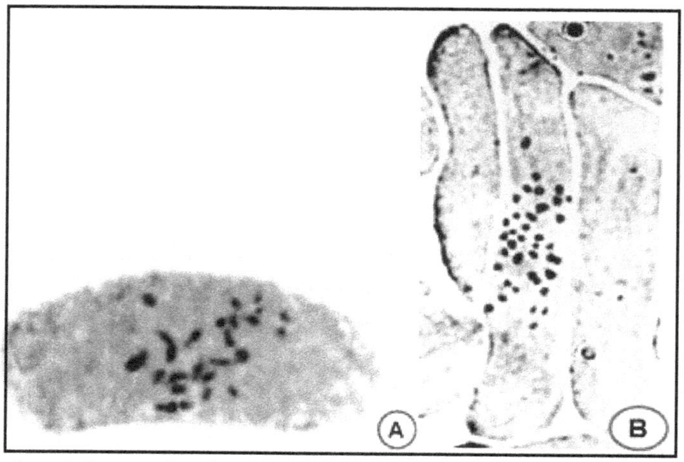

Cells from the root-tips of shoots of endosperm origin: **A.** showing diploid number of chromosomes (2n=2x=24), **B.** triploid number of chromosomes (2n=3x=36).

Concluding Remarks

Like all other plant parts, endosperm can also respond the same way under in vitro conditions irrespective of their genetic constitutions. Hence, it helped in changing the misconception that endosperm being a "dead tissue" has now been contradicted by several reports suggesting the full plant regeneration from endosperm. Despite the success of plant regeneration from endosperm cultures in a number of systems, this protocol for production of triploid plants remained unutilised mostly. It may be basically due to the difficulty in obtaining organogenic callus from the mature or immature endosperms. The ploidy variation exhibited by endosperm derived plantlets is another difficulty which limits this technique. More efforts should be focussed on endosperm regeneration from plants where seedlessness is employed to improve the quality of fruits and plants that has economically useful vegetative parts.

References

- Valster, A. H.; et al. (2000). "Plant GTPases: the Rhos in bloom". Trends in Cell Biology. 10 (4): 141–146. doi:10.1016/s0962-8924(00)01728-1
- Mayer, K. F. X; et al. (1998). "Role of WUSCHEL in Regulating Stem Cell Fate in the Arabidopsis Shoot Meristem". Cell. 95: 805–815. doi:10.1016/S0092-8674(00)81703-1. PMID 9865698
- J. Mark Cock; Sheila McCormick (July 2001). "A Large Family of Genes That Share Homology with CLAVATA3". Plant Physiology. 126: 939–942. doi:10.1104/pp.126.3.939. PMC 1540125. PMID 11457943
- Galun, Esra (2007). Plant Patterning: Structural and Molecular Genetic Aspects. World Scientific Publishing Company. p. 333. ISBN 9789812704085
- "Branching out: new class of plant hormones inhibits branch formation". Nature. 455 (7210). 2008-09-11. Retrieved 2009-04-30

5

Genetic Modification of Plants

Genetically modified plants are those plants which have been genetically altered using various methods like protoplast isolation, somatic hybridization, gene gun method, electroporation, etc. for the purpose of introducing favorable traits. This chapter deals with genetically modified plants and the various methods used in a comprehensive manner.

Genetically modified (GM) plants are also called transgenic plants. They are designed to acquire useful quality attributes such as insect resistance, herbicide tolerance, abiotic stress tolerance, disease resistance, high nutritional quality, high yield potential, delayed ripening, enhanced ornamental value, male sterility, and production of edible vaccines. Another major goal for raising the GM plants is their application as bioreactors for the production of nutraceuticals, therapeutic agents, antigens, monoclonal antibody fragments biopolymers, and so forth. Thus, GM plants can potentially affect many aspects of modern society, including agricultural production and medical treatment. Despite these potential applications, the use of GM plants for human welfare has been restricted owing to various concerns raised by the public and the critics. These concerns are divided into different categories, namely, health, nutritional, environmental, ecological, socioeconomic, and ethical concerns. These concerns include those arising due to properties of GM plants themselves, those resulting from the spread of the transgenes to other organisms, and also those resulting from their release into the environment. Such concerns have led to the withdrawal of commercialization of Bt cotton and Bt brinjal in India.

The campaign against GM plants was fueled by the instances of transgenic potatoes reported to be deleterious to rats, contamination of commercial corn products with unapproved StarLink and killing of monarch butterfly by Bt corn pollen. Furthermore, the nongovernmental organizations (NGOs) such as Gene Campaign, Center for Sustainable Agriculture, Research Foundation for Science, Technology and Ecology, Greenpeace, and Friends of the Earth have also raised concerns related to genetic manipulation of plants. The regulators, activists, media personnel and scientific journals have been undiscriminating and overly tolerant of the misrepresentations and distortions of anti-GM activists. There are not even scientific explanations for some of the concerns, but today the amount of misinformation is such that it has become difficult to separate truth from public perception about the GM plants. The biotechnology scientists, however, believe that GM plants should be given public acceptance because most of the concerns are not specific for GM plants and can exist for non-GM plants as well.

Concerns Related to Health and Nutritional Status

In case the products of GM plants are to be consumed by humans and animals, there is always a fear and risk in the society that these plants may create health problems or may lead to the development of newer microbial strains that may be pathogenic. Further, the plants themselves may be susceptible to such risks. Public and critics are also skeptical about the nutritional content and quality of the GM plants.

Susceptibility to Allergens

One of the major distressing problems with nontraditional proteins in GM foods is the risk of introducing allergens (usually glycoproteins) into the food supply of humans and animals. The public is concerned about the nature of these new food proteins as their allergenic or nonallergenic qualities are unknown. Allergenicity has been demonstrated in transgenic soybeans due to the transfer of a major food allergen from Brazil nuts. On the other hand, the scientists believe that the food allergens are found only in a few defined sources (peanut and other grain legumes, shellfish, tree nuts, etc.), and hence, only a dozen foods may produce allergic reactions. Moreover, allergenicity occurs when these food allergens are present in large proportions in the food and the individuals are sensitized to them over time to cause any adverse effects. Thus, it is highly unlikely for new allergens to be introduced into the food supply from GM plants.

Transfer of Antibiotic Resistance Gene to Microbes and Reduced Efficacy of Antibiotic Therapy

Public is also concerned about the potential risks associated with gene transfer from plants to microbes. It is speculated that the consumption of GM foods containing antibiotic resistance marker gene (e.g., Npt II gene encoding neomycin phosphotransferase for resistance to kanamycin and neomycin or gene encoding β-lactamase for resistance to ampicillin) by humans and animals may lead to transfer of these genes from GM food to microflora in the gut of humans and animals or to the pathogens in the environment transforming them into strains that are resistant to antibiotic therapy.

The transfer of antibiotic resistance gene to unrelated microorganisms such as Aspergillus niger has also been demonstrated. Biotechnology scientists, however, are of the opinion that the Npt II gene used to develop GM plants currently in the market is safe for use because there is no evidence of allergenicity or toxicity related to it. Moreover, humans are also susceptible to consuming several kanamycin resistant bacteria that occur naturally in the environment. Human gut is reported to contain 1012 kanamycin-resistant bacteria and by consuming a tomato harboring Npt II gene, the increase in frequency of kanamycin-resistant bacteria in the gut amounts only to 10–6%. Furthermore, acid conditions prevalent in stomach or rumen inactivate or degrade the encoded enzyme, neomycin phosphotransferase II.

Also neomycin phosphotransferase II requires ATP for its activity, which is present in extremely low concentrations in the gut. Regarding the use of gene for selection of bacterial recombinants, it is not transferred to plants. Moreover, the gene is considered safe because it does not encode for any product in plants. The Npt II and genes have been declared safe to use in GM plants. The public is, however, reluctant to accept this fact. Looking to the views of public, scientists have also developed nonresistance based selectable marker genes such as green fluorescent protein encoding gene (Gfp) and β-glucuronidase gene (Uid A). Besides, intron-containing Npt II gene has also been assessed as an efficient selectable marker in plant transformation. Due to insertion of intron in the Npt II gene, the theoretical risk of gene flow from GM plants to enteric bacteria is eliminated. Strategies for the removal of antibiotic resistance genes have also been devised.

One such strategy is the cloning of selectable marker gene and the transgene on two separate transfer DNA (T-DNA) molecules in a single plasmid or on two separate plasmids that are contained in one or more Agrobacterium tumefaciens strains used for plant transformation. The transgene and selectable marker gene are, thus, inserted at the loci, which should recombine at reasonably high frequencies so that the transgene can be segregated from the selectable marker gene in the next generation. Second strategy to eliminate the selectable marker gene is to flank it with direct repeats of recognition sites for a site-specific recombinase so that the marker gene can be easily excised from the plant genome by recombinase-mediated site-specific recombination. Examples included in this category are the Cre/lox recombination system of bacteriophage P1, Flp/frt recombination system of yeast 2 μm plasmid and R/Rs system of Zygosaccharomyces rouxii. A common feature of these systems is that the first round of transformation produces transgenic plants with the selection marker between two directly oriented recognition sites for the respective recombinase. After expression of recombinase, either by crossing in plants expressing the enzyme, by transient expression via second transformation, or by the use of an inducible promoter, the recombinase reaction is initiated resulting in marker-free transgenic plants. Marker gene may also be eliminated by placing it on a transposable element resulting in its loss after transposition.

The transgene by itself may be mobile and the activation of transposase allows the relocation of the desired transgene to a new chromosomal position. Genetic crosses and/or segregation may dissociate the two transgenes. Another novel strategy for the production of marker-free GM plants involves DNA deletion based on intrachromosomal homologous recombination between two homologous sequences, for example, by incorporating att sequence of λ bacteriophage.

Development of New-Line Microbial Strains

The third health risk is related to the ability of GM plants to create new toxic organisms. It is speculated that some nonpest microbial strains may acquire pathogenic trait by gene flow from GM plants. The risk can also be a new host being infected by a virus or recombining to form a more deadly virulent virus. Some plant pathologists also hypothesize that development of virus-resistant plants may allow viruses to infect new hosts through transencapsidation. Virus-resistant plants may also lead to the creation of new viruses through an exchange of genetic material or recombination between RNA virus genomes. Another matter of concern is that a small fraction of the DNA released from GM plants into soil may bind to the clay particles and hence protected from degradation. It is speculated that the soil bacteria may undergo transformation with the exogenous DNA of GM plant. This is, however, a rare possibility as the amount of DNA derived from GM plants as a proportion of the total DNA in the soil is likely to be very small, even if such plants are grown on a commercial scale. Moreover, the longevity of DNA in soil depends on various factors, including soil type and the presence of deoxyribonucleases in soil. target DNA from transgenic tobacco plants gets degraded within 40 days.

Skepticism About Nutritional Status

Critics of GM crops have raised various concerns about the potential of golden rice to combat vitamin A deficiency (VAD). The primary concern amongst these is the presence of insufficient vitamin A in golden rice. There are still doubts about the speed of degradation of vitamin A after harvesting the plant and the amount of vitamin A left after cooking. Vandana Shiva, an Indian anti-GMO activist, has criticized golden rice by saying "the golden rice is a hoax," "golden rice is a blind approach for blindness control," and "golden rice is just a recipe for creating hunger and malnutrition". She argues that the golden rice fails to pass the vitamin A need test and is incapable of removing VAD. It is calculated that one serving contains 30 g of rice on dry weight basis and golden rice can provide only 9.9 µg of vitamin A, that is, only 1.32% of the required daily allowance (RDA) of 750 µg. Even with the daily consumption of 100 g golden rice, only 4.4% of the required daily allowance will be met. Thus, an adult has to consume 2 kg 272 g of golden rice per day to complete his daily requirements of vitamin A. She is also of the view that, besides creating VAD, golden rice will also create deficiency in other micronutrients and nutrients.

This is because the raw milled rice has a low content of fat (0.5 g/100 g), which is necessary for vitamin A uptake, low content of protein (6.8 g/100 g), which is required as a carrier molecule, and low content of iron (0.7 g/100 g), which is required for the conversion of β-carotene to vitamin A. Friends of the Earth, Greenpeace, and Vandana Shiva further emphasize that there is no need of golden rice to combat VAD as superior alternatives such as sweet potato, green leafy vegetables, coriander, amaranth, carrot, pumpkin, mango, jackfruit exist in nature. It is reported that certain underutilized plants also exhibit far more nutritional value (vitamin A and other nutrients) than golden rice, for example, a combination of rice and leaves of Moringa (drumstick) tree, a native to India. Similarly, in contrast to rice, amaranth grain contains forty times more calcium, four times iron, and twice as much protein.

The ragi millet, grown in India, has thirty five times more calcium than rice, twice as much iron, and five times more minerals. It is opined that golden rice is not capable of increasing the production of β-carotene. Even if the target of 33.3 µg of vitamin A in 100 g of rice is achieved, it will be only 2.8% of β-carotene that can be obtained from amaranth leaves, 2.4% as that obtained from coriander leaves, curry leaves, and drumstick leaves. Thus, a far more efficient route to removing VAD is biodiversity conservation and propagation of naturally occurring vitamin A rich plants (wild-type or underutilized) in agriculture and diets. Even the World Bank has admitted that rediscovering and use of local plants and conservation of vitamin A rich green leafy vegetables and fruits have dramatically reduced VAD threatened children over the past 20 years in very cheap and efficient ways.

It is also speculated that the cultivation of golden rice will lead to major water scarcity since it is a water intensive crop and displaces water prudent sources of vitamin A. The scientists, on the other hand, believe that the traditional breeding methods have been unsuccessful in producing crops containing a high vitamin A concentration and most national authorities rely on expensive and complicated supplementation programs to address the problem. They also believe that a varied diet is beyond the means of many of the poor and they have to rely on one or few foods to provide complete nutrition, for example, rice.

Thus, golden rice may be a useful tool to help treat the problem of VAD in young children living in the tropics. They also emphasize that the critics are ignoring the fact that VAD disorders result from a deficiency of vitamin A and not its complete absence in the diet and the VAD individuals lack only 10%–50% of their daily requirements. Hence, any additional contribution toward daily requirements would be useful. In 2005, a team of researchers at Syngenta have produced a variety of golden rice, called "Golden rice 2," which produces twenty-three times more carotenoids than golden rice (up to 37μg/g) and preferentially accumulates β-carotene (up to 31μg/g of the 37μg/g of carotenoids). The Rockefeller Foundation emphasized that the new strains of golden rice contain substantially higher levels of β-carotene than the early versions on which the opponents based their calculations. In order to meet the RDA, 144g of the most high-yielding golden rice strains would have to be eaten.

Environmental and Ecological Concerns

Large-scale cultivation of GM plants expressing viral and bacterial genes and their release into the environment is considered to be a threat and called as "genetic pollution" by the critics. The risk of a transgene spreading in the environment is related to the likelihood for out-crossing, horizontal gene transfer, and the phenotype imparted by the gene. Debates about the commercial introduction of GM plants in some parts of the world have led to questions about their potential impact on the environment unless necessary safeguards are taken into account. It should, however, be acknowledged that agriculture inevitably has an impact on the environment and these concerns are not specific for GM plants.

Transgene Escape to Wild-Type Plants

There is a potential risk that the GM plants may hybridize (or cross-breed) with sexually compatible wild-type species. This genetic exchange is possible due to wind pollination, biotic pollination or seed dispersal. This may have an impact on the environment through the production of hybrids and their progeny. In an example, virus-resistant squash commercialized in 1994 was demonstrated to transfer its virus resistance gene to wild squash (Cucurbita pepo), an agricultural weed native to the southern United States, thereby decreasing its value to squash breeders. On the other hand, it is significant to note that for an effective pollen transfer to occur, the GM plants must be close enough to the wild species, should flower at same time, and must be genetically compatible. Further, the risk of any gene transfer to related weedy species through pollen has been eliminated by devising chloroplast transformation procedures. This is because, in many crop species, chloroplasts display only maternal inheritance.

Selective Advantage to GM Plants in Natural Environments and Generation of Super weeds

The concern of gene flow from GM plants to weedy relatives via pollination is quite intense. It is considered that the transfer of encoded characteristics to weed species could potentially give them a selective advantage, consequently leading to the generation of "superweeds." Moreover, the newly introduced traits may make a plant, especially herbicide tolerant plant, more persistent or invasive (weedy) in agricultural habitats. It is, however, pertinent to note that the risk of gene transfer to weeds is similar with both conventional and GM plants and is not contingent on how these genes have been introduced into plants. Such a risk of gene flow has always existed since the advent of modern plant breeding, even when there were no GM plants, and this can occur where possible. Several studies have demonstrated that tolerance to particular herbicide is often more likely to develop by evolution from within the weed gene pool rather than by gene flow from herbicide-tolerant plants. Nevertheless, the current scientific evidence indicates that the weediness arises from many different characters and that the addition of one gene is unlikely to cause a crop to become a weed. The transfer of novel genes from transgenics (or even conventionally bred plants) to weeds depends on the nature of the novel gene and the biology and ecology of the recipient weed species. The probability of successful out-crossing thus depends on sexual compatibility, physical proximity, distance of pollen movement both out of and into the GM plants, and ecology of recipient species. Thus, only a few plants such as oilseed rape, barley, wheat, beans, and sugar beet can hybridize with weeds. For example, oilseed rape has been reported to hybridize with hoary mustard, wild radish, and other wild Brassica species. Furthermore, the transfer of herbicide tolerance gene is unlikely to confer any competitive advantage to hybrids outside agricultural areas. It is also comforting to recognize that there is no proven evidence of enhanced persistence or invasiveness of GM plants and no major super weeds have developed so far.

Effect on Nutritional Composition of Plants

It is also speculated that the nutritional composition of GM products may be affected in GM plants. Another concern is that the transgenes from animals (obtained from fishes, mouse, human, and microbes) introduced into GM plant for molecular farming may pose a risk of changing the fundamental nature of vegetables. In a study, it was reported that as compared to non-GM soybean, GM plants exhibited lower levels of isoflavones . This finding also raised a doubt on the regulatory system for the release of the GM plants. However, later it was found that the concentration of isoflavone in GM soybean was within the normal range.

Mixing Genes from Unrelated Species (Interbreeding)

The public is worried about the risk that the GM plants can spread through nature and interbreed with natural organisms, thereby contaminating "non-GM" environments. This would in turn affect the future generations in an unforeseeable and uncontrollable way . Such worries, however, ignore the history of plant breeding and the existing overwhelming sequence similarity of genes across kingdoms.

Development of Tolerance to Target Herbicide

It is viewed that the repeated use of the same herbicide in the same area to remove weeds amongst genetically modified herbicide-resistant crops (HRCs) (tolerant to single herbicide) will exacerbate the problem of herbicide-tolerant weeds . Another matter of concern relates to the plants carrying different herbicide tolerance genes to become multiply tolerant to several herbicides by pollination between adjacent plants . In several closely studied examples in Canada, farmers have detected oilseed rape plants tolerant to three different herbicides (note that two were acquired from GM plants and the third possibly from conventional breeding) . The development of multiple tolerances in "volunteer" crop plants (from seeds remaining viable in agricultural soil) may also exert an impact on the environment by necessitating the use of less environment-friendly and possibly outdated herbicides by the farmers. On the other hand, the proponents believe that herbicide resistance develops due to excessive application of herbicide and is not exclusively associated with gene transfer from genetically modified HRCs. Thus, the pressure on weeds to evolve resistant biotypes has been reported to be pronounced with the excessive application of herbicides such as glyphosate, sulphonylureas, and imidazolinones.

Sustainable Resistance in Insect Pests

It is possible that the widespread use of disease-resistant GM plants may lead to the evolution of several insect pests that are resistant to pesticides. For example, Bt crops may develop resistance to Bt biopesticide, a permitted biopesticide successfully used by organic farmers in the integrated pest management (IPM) programs. There is to date no reported evidence of insect resistance to Bt crops under field conditions although Bt resistant insects (e.g., cotton budworm and bollworm) have been observed in areas where Bt biopesticides are sprayed on crops . It has been a matter of concern that the development of such resistance may lead to the loss of the potential of the Bt biopesticide, which may in turn make it necessary for organic farmers to resort to less environmentally acceptable chemical pesticides. Therefore, proper resistance management strategies along with this comparatively newer technology are imperative. The most widely used is the 'high-dose refuge' strategy designed to prevent or delay the emergence of Bt toxin-resistant insects. Scientists are of the opinion that this strategy should be followed without fail, as the rate of noncompliance can increase the risk of plant resistance breakdown.

Harm to Nontarget Organisms

Nontarget effect, that is, undesirable effect of a novel gene (usually conferring pest or disease resistance) on "friendly" organisms in the environment, is another concern related to GM plants . As many nontarget microbes harbor on plant surfaces or some insects harbor on flowers, it becomes quite challenging to target the insect resistance gene product to appropriate plant tissues and hence kill pests without exerting any adverse effect on friendly organisms such as pollinators and biological control agents. This is particularly difficult where the benign or beneficial organism is related and physiologically similar to the pest to be targeted. One of the most significant studies of nontarget impacts of GM plants has been the killing of monarch butterfly in the United States by Bt insecticidal proteins. It should, however, be noted that the pesticidal sprays used on Bt or non-Bt corn may be more harmful to the monarch butterfly as compared to Bt corn pollen. Thus, in evaluating the use of Bt crops and the possible environmental damage caused, it is important to take into account the environmental damage caused by the use of pesticides in agriculture generally. It is argued that millions of birds and billions of insects, both harmful and beneficial, are killed each year due to excessive use of pesticides. It is, however, suggested that the scale and pattern of use may mitigate the effects of Bt on nontarget populations. Furthermore, when toxins are produced within plant tissues, nontarget organisms are exposed to a much lesser extent than with spray applications because only those organisms which feed on the plant tissues come into contact with the toxin.

Harmful effect of Bt toxin residues in the soil after harvest of the GM crop on soil invertebrates has been another matter of concern. An investigation of the effect of Cry1Ab released from the roots and crop residues on soil organisms revealed the presence of toxin in the guts and casts of tested earthworms. There was, however, no significant difference in their mortality or weight. Moreover, no difference in the total number of other soil organisms (including nematodes, protozoa, bacteria, and fungi) between the soil rhizosphere of Bt and non-Bt crops was detected.

Increased Use of Chemicals in Agriculture

On one hand, the transgenes conferring herbicide resistance have been criticized because these would maintain, if not promote, the use of herbicides and their attendant problems [125, 126]. Similarly, there is a concern that the insect-resistant and disease-resistant GM plants will increase the application of insecticides and pesticides, respectively. On the contrary, reports demonstrate that there is no significant change in the overall amount of herbicide use in the United States since the introduction of GM soybeans . An analysis by soybean growers at the United States has shown that $7.2 millions of other herbicides were replaced by $5.4 millions of glyphosate . This substitution, thus, resulted in the replacement of highly toxic and more persistent herbicides with that of glyphosate. Furthermore, it has been reported that herbicide-tolerant oilseed rape eliminates the use of >6,000 tons of herbicide in the growing season .

Loss of Biodiversity

The public has long been worried about the loss of plant biodiversity due to global industrialization, urbanization, and the popularity of conventionally-bred high-yielding varieties. It is speculated that the biodiversity will be further threatened due to the encouraging use of GM plants. This is because development of GM plants may favor monocultures, that is, plants of a single kind, which are best suitable for one or other conditions or produce one product. Further, the transformation of more natural ecosystems into agricultural lands for planting GM plants is adding to this ecological instability.

Another point of concern is the loss of weed diversity that may occur due to gene flow from HRCs to weeds . It is argued that because the currently available HRCs confer tolerance to broad-spectrum herbicides such as glufosinate and glyphosate, their extensive use may shift the diversity of weeds in agricultural habitats. However, weeds exhibit considerable plasticity and adapt to a wide range of cultivation practices. Experience with conventional agriculture has shown that weed species composition varies within the same crop among different fields and at different times of year. Thus, weed population shifts are natural ecological phenomena in crop management and should not be viewed as exclusive to GM plants.

Unpredictable Gene Expression

It is speculated that the random gene insertion, transgene instability, and genomic disruption due to gene transfer may result in unpredictable gene expression. Such a risk is, however, unlikely to be unique to GM plants or of any significance considering our current knowledge of genomic flux in plants.

Alteration in Evolutionary Pattern

Plants adapt to the fluctuations in the environment through changing their genes and developing better races called "evolved races." These mutations, however, occur at a very low frequency (i.e., one in about 109/gene/generation). It is hypothesized that the cultivation of GM plants by the farmers at an increasing rate throughout the world may change the evolutionary pattern drastically . Another concern is the evolution of non-GM plants through hybridization with GM plants.

Loss of Ecosystem in Marginal Lands

As new plants are introduced mainly to marginal lands, loss of natural ecosystems in these areas has also been a matter of concern.

Contamination of Soil and Water

It is also sometimes argued that the widespread introduction of HRCs will increase the use of herbicides, which will in turn contribute to the contamination of soil and ground water. However, this is not the case. The cultivation of HRCs in the United States has been reported to facilitate zero-till agronomic system, which contributes to a reduction in soil erosion. The release of Bt toxin into the soil after harvest of Bt crops is also viewed as a risk factor associated with the cultivation of Bt crops [123, 124, 130]. It has been found that Bttoxins remain active in soil; however, it is not necessarily an environmental hazard because Bt toxins must be ingested and affect only selected groups of insects. Moreover, the potential leaching rate of Bt toxin is reduced due to its binding and adsorption on clay particles.

Genetically Modified Crops

The term genetically modified (GM), as it is commonly used, refers to the transfer of genes between organisms using a series of laboratory techniques for cloning genes, splicing DNA segments together, and inserting genes into cells. Collectively, these techniques are known as recombinant DNA technology. Other terms used for GM plants or foods derived from them are genetically modified organism (GMO), genetically engineered (GE), bioengineered, and transgenic. 'Genetically modified' is an imprecise term and a potentially confusing one, in that virtually everything we eat has been modified genetically through domestication from wild species and many generations of selection by humans for desirable traits. The term is used here because it is the one most widely used to indicate the use of recombinant DNA technology.

Traits that have been Modified in GM Crops

Insect-resistant crops contain genes from the soil bacterium Bacillus thuringiensis (Bt). The protein produced in the plant by the Bt gene is toxic to a targeted group of insects—for example European corn borer or corn rootworm—but not to mammals. The most common herbicide tolerant (HT) crops are known as Roundup Ready, meaning they are tolerant to glyphosate (the active ingredient in Roundup herbicide). Glyphosate inactivates a key enzyme involved in amino acid synthesis that is present in all green plants; therefore, it is an effective broad spectrum herbicide against nearly all weeds. Roundup Ready crops have been engineered to produce a resistant form of the enzyme, so they remain healthy even after being sprayed with glyphosate. Some cultivars of corn and cotton are referred to as 'stacked', meaning they have transgenes for both insect resistance and HT.

Potential GM Crops of the Future

Some potential applications of GM crop technology are:

- Nutritional enhancement: Higher vitamin content; more healthful fatty acid profiles;
- Stress tolerance: Tolerance to high and low temperatures, salinity, and drought;
- Disease resistance: For example, orange trees resistant to citrus greening disease or American chestnut trees resistant to fungal blight;
- Biofuels: Plants with altered cell wall composition for more efficient conversion to ethanol;
- Phytoremediation: Plants that extract and concentrate contaminants like heavy metals from polluted sites.

Difference between GM Technology and Plant Breeding Techniques

The era of scientific crop improvement dates back to around 1900, when the impact of Gregor Mendel's studies on trait inheritance in peas became widely recognized. Since then, a broad range of techniques has been developed to improve crop yields, quality, and resistance to disease, insects, and environmental stress. Most plant breeding programs rely on manual cross-pollination between genetically distinct plants to create new combinations of genes. The progeny plants are intensively evaluated over several generations and the best ones are selected for potential release as new varieties. An example is a tomato variety that is selected for disease resistance and tolerance to cool temperatures. Other techniques included within the conventional plant breeding toolbox are development of hybrid varieties by crossing two parental strains to produce offspring with increased vigor; and induced mutations to create useful variation. GM technology is much more precise in that it transfers only the desired gene or genes to the recipient plant. Another branch of agricultural biotechnology—distinct from GM technology—involves selecting plants for DNA patterns known to be associated with favorable traits such as higher yield or disease resistance.

The Shared DNA Code

Most organisms store their genetic information in the form of DNA molecules in chromosomes. The sequence of chemical bases in a DNA strand encodes a specific order of amino acids, which are the building blocks of proteins. Proteins carry out many functions in cells and tissues, which together are responsible for an organism's characteristics. Because most life forms share this same language of heredity—and due to scientific advances in molecular biology—it is now possible to transfer a gene from one species to another, for example from a bacterium to a plant, and have it function in its new host.

Types of Genetically Modified Plants

Transgenic Plants

Transgenic plants or genetically modified plants are plants whose DNA is modified using genetic engineering techniques. In most cases the aim is to introduce a new trait to the plant which does not occur naturally in this species. Examples include resistance to certain pests, diseases or environmental conditions, or the production of a certain nutrient or pharmaceutical agent.

A transgenic crop plant contains a gene or genes which have been artificially inserted instead of the plant acquiring them through pollination. The inserted gene sequence, known as the transgene, may come from another unrelated plant or from a completely different species. Plants containing transgenes are often called genetically modified or GM crops.

Reasons for Making Transgenic Crop Plants

- The process of transgenic plant development primarily aims at assembling a combination of genes in a crop plant which will make it as useful and productive as possible. Depending on where and for what purpose the plant is grown, desirable genes may provide features such as higher yield or improved quality, pest or disease resistance, or tolerance to heat, cold and drought.
- Combining the best genes in one plant is a long and difficult process, especially as traditional plant breeding has been limited to artificially crossing plants within the same species or with closely related species to bring different genes together.

  For example, a gene for protein in soybean could not be transferred to a completely different crop such as corn using traditional techniques. Transgenic technology enables plant breeders to bring together in one plant useful genes from a wide range of living sources, not just from within the crop species or from closely related plants.
- This technology provides the means for identifying and isolating genes controlling specific characteristics in one kind of organism, and for moving copies of those genes into another quite different organism, which will then also have those characteristics.
- This powerful tool enables plant breeders to do what they have always done-to generate more useful and productive crop varieties containing new combinations of genes-but it expands the possibilities beyond the limitations imposed by traditional cross-pollination and selection techniques.

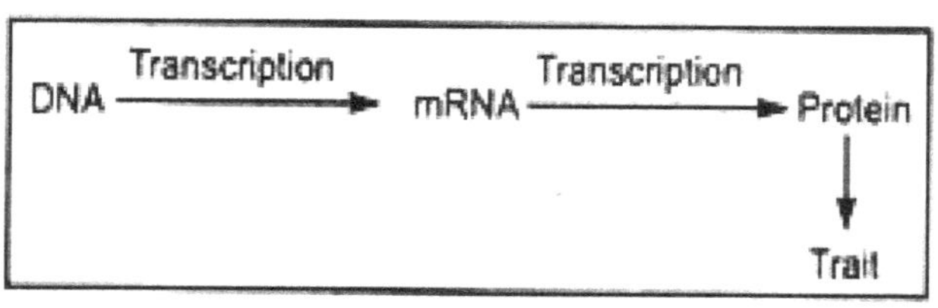

Fundamentals of Transgenic Plant Development

The underlying reason that transgenic plants can be constructed is the universal presence of DNA (deoxyribonucleic acid) in the cells of all living organisms. This molecule stores the organism's genetic information and orchestrates the metabolic processes of life. Genes are discrete segments of DNA that encode the information necessary for assembly of a specific protein.

A specific protein (or an enzyme) encodes for a particular trait. In the production of a transgenic plant our primary aim is to transfer a foreign gene, encoding for some novel traits, into the genome of the plant stably.

After transferring we also need the transgene to integrate and express in the plant's cells. This process as a whole generates a new variety of plant which is new in its own kind and interests us in its massive large scale cultivation.

Steps Involved in the Production of Transgenic Plants

The fundamental steps involved in the transgenic plant production are as follows:

Step: Identifying, Isolation and Cloning of Genes for Agriculturally Important Traits

The very first step in the generation of transgenic plant is to identify and isolate the novel transgene that we want to transfer into the genome of the target plant. Usually, identifying a single gene involved with a trait is not sufficient. We also have to understand how the gene is regulated, what other effects it might have on the plant, and how it interacts with other genes active in the same biochemical pathway.

Step: Designing Gene Construct for Insertion

After entering the plant cell the transgene must inter-grate into the genome of the plant stably and express itself successfully so as to produce higher amount of transgenic protein which will be indirectly reflected in the trait controlled by it.

To achieve this we have to design a "gene construct" or gene-set, having all the DNA segments necessary to achieve the integration and expression of the transgene. Once a gene has been isolated and cloned (amplified in a bacterial vector), it must undergo several modifications before it can be effectively inserted into a plant.

A gene-set which will be transferred to the target plant has following segments:

A Promoter Sequence

This must be added for the gene to be correctly expressed (i.e., translated into a protein product). The promoter is the on/off switch that controls when and where in the plant the gene will be expressed. To date, most promoters in transgenic crop varieties have been "constitutive", i.e., causing gene expression throughout the life cycle of the plant in most tissues.

The most commonly used constitutive promoter is CaMV 35S, from the cauliflower mosaic virus, which generally results in a high degree of expression in plants. Other promoters are more specific and respond to cues in the plant's internal or external environment. An example of a light-inducible promoteris the promoter from the cab gene, encoding the major chlorophyll a/b binding protein.

The Transgene

Sometimes, the transgene is modified to achieve greater expression in a plant.

For example, the Bt gene for insect resistance is of bacterial origin and has a higher percentage of A-T nucleotide pairs compared to plants, which prefer G-C nucleotide pairs. In a clever modification, researchers substituted A-T nucleotides with G-C nucleotides in the Bt gene without significantly changing the amino acid sequence. The result was enhanced production of the gene product in plant cells.

Termination Sequence

This signals to the cellular machinery that the end of the gene sequence has been reached.

A Selectable Marker Gene

This is added in order to identify plant cells or tissues that have successfully integrated the transgene. This is necessary because achieving incorporation and expression of transgenes in plant cells is a rare event, occurring in just a few per cent of the targeted tissues or cells.

Selectable marker genes encode proteins that provide resistance to agents that are normally toxic to plants, such as antibiotics or herbicides. Only plant cells that have integrated the selectable marker gene will survive when grown on a medium containing the appropriate antibiotic or herbicide.

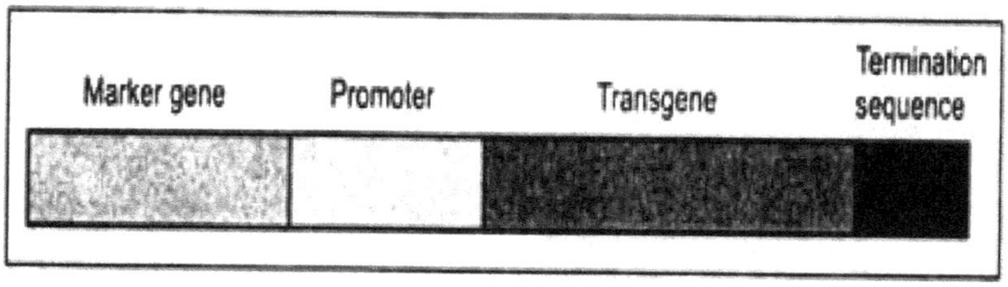

The essential features of an ideal reporter gene are:

1. Lack of endogenous activity in plant cells of the concerned enzyme,
2. An efficient and easy detection,
3. A relatively rapid degradation of the enzyme.

The commonly used selectable marker genes include those conferring resistance to the antibiotics kanamycin (nptII, encoding neomycin phosphotransferase) and hygromycin (hptIV, encoding hygromycin phosphotransferase, isolated from E. coli); and broad range herbicides glyphosate (modified versions of the enzyme EPSPS, 5-enolpyruvate shikimate-3-phosphate synthase, isolated from E. coli or Salmonella typhimurium), phosphinothricin (bar, isolated from Streptomyces hygroscopicus, codes for phophinothricin acetyltransferase), etc.

Step: Transforming Target Plants with the Gene Construct

There are two ways of genetically transforming the target plant:

1. Vector mediated gene transfer,
2. Vector less or direct gene transfer.

Step: Selection of The Transgenic Plant Tissue/Cells

Following the gene insertion process, plant tissues are transferred to a selective medium containing an antibiotic or herbicide, depending on which selectable marker was used. Only plants expressing the selectable marker gene will survive and it is assumed that these plants will also possess the transgene of interest.

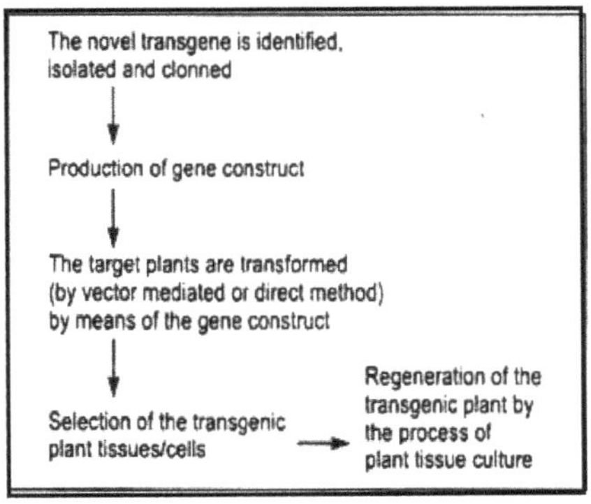

Step: Regeneration of the Transgenic Plants

To obtain whole plants from transgenic tissues, they are grown under controlled environmental conditions in a series of media containing nutrients and hormones by the process of plant tissue culture.

Promoters used in construction of gene construct

Promoter	Source	Relative Activity
35S	CaMV 35S RNA gene	Constitutive, high activity; most commonly used in dicots
35S + Adh1-I 1	35S promoter + first intron of maize Adh1 gene	Enhanced promoter activity; constitutive
35S + sh1-I 1	35S promoter + first intron of maize shrunken 1 gene	Better than 35S + Adh1-I 1 in monocots; constitutive
Adh1	Promoter of alcohol dehydrogenase gene of maize	Moderate activity in cereals; anaerobic expression
Emu	Modified from Adh1 promoter and its first intron	Moderate activity in cereals; anaerobic expression
Act1 + Act-I 1	Rice actin gene + its first intron	Moderate activity; constitutive
Ubi1 + Ubi1-I 1 (or I 6)	Maize ubiquitin 1 gene promoter + its first (or sixth) intron	High activity in cereals; constitutive
Vicilin promoter	Pea vicilin storage protein gene	Seed-specific promoter

Integration of the Transgene in the Genome of the Target Plant

In general, transgenes integrate at random sites in any of the chromosomes of the genome of host cells. Usually, in a given cell, integration occurs at a single location. As a result, different cells may be expected to show integration of the transgene at different chromosomal locations.

The number of copies integrated per genome ranges from one to several hundred. In general, multiple copies are integrated when large amounts of DNA are used for transfection, while single copies are integrated with smaller amounts.

When multiple copies are integrated, they are mostly integrated at one site joined to each other head-to-tail, i.e., as a concatemer. However, in a small proportion of cases, the multiple copies are located at several sites in the same genome.

The mechanism of random integration is not known. The entire gene construct, including the vector DNA, becomes integrated. When two different gene constructs are mixed and used for transfection, they tend to be integrated together at the same site; this is known as co-transfection. The sequences flanking a gene on either side influence the expression of this gene.

Therefore, the same transgene integrated at different locations in the genome may show different levels of expression; this is known as position effect. Transgene integration frequently leads to various forms of rearrangements, e.g., duplication, deletion, etc., near the site of integration.

If these changes are large enough, the host gene located at the site of integration may become non-functional. A host gene would also become non-functional if the transgene becomes integrated within the coding region of this gene. When integration of a transgene leads to the loss of function of a host gene, it is called insertional mutagenesis; it often produces aberrant phenotypes.

Analysis of Transgene Integration

The integration of transgene into the genome is confirmed by Southern hybridization of genomic DNA extracted from the considered transgenic individuals. The DNA is digested with a suitable restriction enzyme prior to electrophoresis.

By choosing appropriate restriction enzymes for DNA digestion, not only the integration of transgene can be established beyond doubt, but information on the number of copies per cell, the orientations of tandemly arranged copies and the presence of single or multiple integration sites is also obtained from Southern hybridization. All the individuals that give positive result with Southern hybridization are regarded as confirmed transgenic.

Detection of mRNA Expression

The mRNAs produced by transgenes is most readily detected if they are with unique sequences, which have no counterparts among those produced by the host genome. A high purity RNA preparation is obtained from the appropriate tissue of transgenic individuals, and is subjected to RNA dot blot hybridization with a radioactive probe specific for the transgene.

Alternatively, the RNA preparation may be used for northern hybridization, which provides additional information on transcript size as well.

Inheritance of Transgenes

The transgenes which are stably integrated are inherited in a Mendelian fashion. They are usually dominant. Instability may occur due to point mutation, like methylation, or rearrangements of the T-DNA region. In addition, homologous recombination between copies of the transgene inserted in the same nucleus can also lead to instability of the gene.

Future Development of Transgenic Technology

New techniques for producing transgenic plants will improve the efficiency of the process and will help resolve some of the environmental and health concerns.

Among the expected changes are the following:

- More efficient transformation, that is, a higher percentage of plant cells will successfully incorporate the transgene.
- Better marker genes to replace the use of antibiotic resistance genes.
- Better control of gene expression through more specific promoters, so that the inserted gene will be active only when and where needed.
- Transfer of multi-gene DNA fragments to modify more complex traits.

Cisgenic Plants

Cisgenic plants are made using genes found within the same species or a closely related one, where conventional plant breeding can occur. Some breeders and scientists argue that cisgenic modification is useful for plants that are difficult to crossbreed by conventional means (such as potatoes), and that plants in the cisgenic category should not require the same regulatory scrutiny as transgenics.

Cisgenesis is the production of genetically modified crops using a donor DNA fragment from the species itself or from a cross compatible species. The newly introduced gene is unchanged and includes its own introns and regulatory sequences and is free of any vector DNA, except T- DNA border sequences that flank the cisgene. The resultant phenotype of the cisgenic plant can be achieved through conventional breeding also; but it will take a much longer time. One of the most important points of cisgenesis is that it introduces only the desired gene, thus avoiding linkage drag that can be resulted from conventional cross breeding and also it eliminates hectic and time consuming backcrossing to recover the recurrent parent genotype. The concept of cisgenesis was introduced by Dutch researchers; Schouten, Krens and Jacobsen in the year 2006.

The worthiness of the GM techniques for developing highly reliable and good quality food products to the world has been set off by public worries about the safety of the derived food and there subsequent finished products. Most particularly, the controversy has spotlighted on the probable unpredictable hazards arising from the agglomeration of certain new substances in crop plants that confers toxicity, allergy and genetic threats to humans.

Benefits of Cisgenesis

- Cisgenesis is as safe as conventional breeding.
- To overcome the problem of linkage drag.
- Original genetic makeup of a crop is preserved any one or few genes are added.
- Escape of foreign genes via pollen grains to natural vegetation is not a problem in cisgenesis.
- In cisgenesis, only the desired genes are introduced without the undesirable genes unlike conventional breeding.

Drawbacks of Cisgenesis

- The gene outside the sexually compatible gene pool cannot be introduced.
- There is also a chance that the introduction of cisgene may influence the expression of genes that are already present in the recipient genome.
- Position effect may lead to alteration of the gene expression and phenotypic differences.
- The production of marker-free plants often requires the implementation or development of new techniques and such techniques may not be readily available.

Advantages of Cisgenesis over Conventional Breeding

1. Removes linkage drag: Introgression of new traits into the cultivated varieties by conventional methods comprises of wide crosses and repeated backcrossing. However, these traits are constantly linked with a large share of unwanted chromosomal region, the so called linkage drag. Some of these genes affect the normal features of the crop as they may engage in the production of diverse kinds of toxins or allergens. Hence, direct transfer of desired genes through cisgenesis into an existing variety without altering any of the properties enviable for the consumers can be accomplished.

2. Maintains the original genetic make-up of the plant varieties: In a hybridization method, the genetic makeup of the progeny plant varies from its parents because it has been a mixture of both the parental genomes. In spite of this, there is a necessity to conserve some part of the genome which reveals certain constructive traits. Through conventional plant breeding such an approach is not possible entirely due to self incompatibility among the vegetatively propagated plants. Cisgenic breeding tools are used in order to get up to four different resistance genes into one variety without changing the other traits of the modified variety.

3. Reduction in pesticide application: The key purpose of cisgenesis is to transfer disease resistance genes to susceptible varieties. The vital goal here is to lessen substantial pesticide application. As a result, there is decline in the input costs of the farmers and decreased pesticide leftovers on the plants and also in their products, which is mostly favoured by the consumers. This reduces the environmental pollution by pesticides and in turn helpes in sustainable agricultural development.

4. Time Saving: In conventional hybridization programmes, there is linkage drag, where there is inheritance of unwanted genes to the progeny. Several backcrossed generations are required to get rid of such kind of undesired genes. Cisgenesis overcomes the problem of linkage drag and only the gene of interest is introduced into the genome of the recipient plant within a short period of time. Thus, this saves a lot of time.

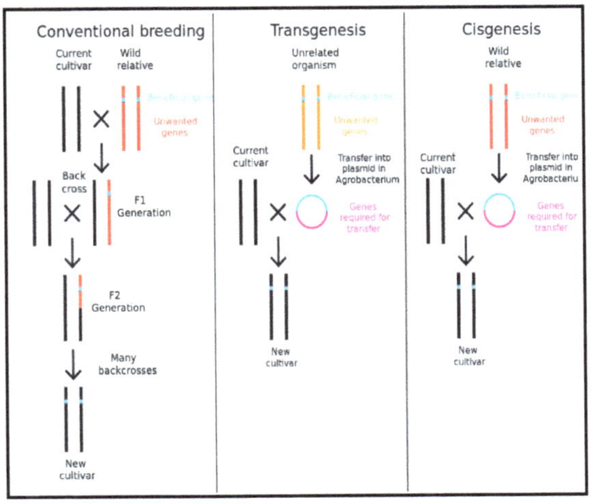

Subgenic

Genetically modified plants can also be developed using gene knockdown or gene knockout to alter the genetic makeup of a plant without incorporating genes from other plants. In 2014, Chinese researcher Gao Caixia filed patents on the creation of a strain of wheat that is resistant to powdery mildew. The strain lacks genes that encode proteins that repress defenses against the mildew. The researchers deleted all three copies of the genes from wheat's hexaploid genome. Gao used the TALENs and CRISPR gene editing tools without adding or changing any other genes. No field trials were immediately planned. The CRISPR technique has also been used to modify white button mushrooms (Agaricus bisporus).

Multiple Trait Integration

From a breeding standpoint, multiple trait integration (MTI) is a four-step process of converting an elite variety/hybrid for value-added traits (e.g. transgenic events) using backcross breeding, ultimately regaining the performance attributes of the target hybrid along with reliable expression of the value-added traits.

Protoplast Isolation

Each plant cell has a definite cellulose wall which is attached with each other by middle lamella made up of pectin. Protoplast is the content which is surrounded only by plasma-lemma. So to get the isolated protoplast it is essential to remove the pectin material to obtain the single cells and then to remove cell wall by enzymatic method.

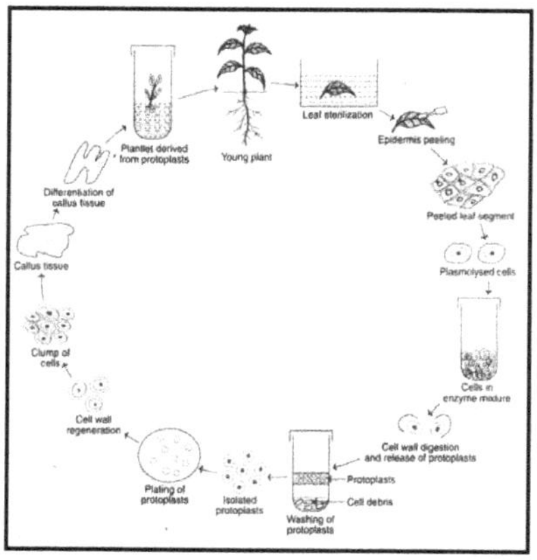

Sources of Explant for Protoplast Isolation

Protoplasts can be isolated directly from the different parts of whole plant which bears the soft parenchymatous tissue (e.g., young fully expanded soft leaves) or indirectly from the in vitro grown plant tissue (e.g., callus tissue). Protoplast yield and viability depends greatly on the type of tissue material from which it is isolated and the method used. Cell suspension cultures may also provide a very good source of protoplast.

Before isolation of protoplast, the source material if it is from in vivo grown plant then it should be properly surface sterilized using the proper method of sterilization. Then any of the methods either mechanical or enzymatic can be used to isolate the protoplast.

Methods of Protoplast Isolation

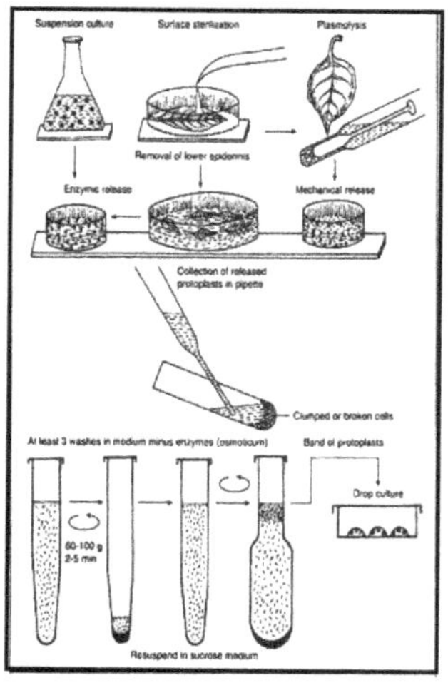

The essential step of protoplast isolation is the proper use of osmoticum. The plant protoplast is an osmotic system where the plasma membrane acts as the semipermeable membrane which equilibrates the outward and inward flow of water. If the inward pressure is more than due to heavy inflow of water the protoplast will burst.

So during protoplast isolation the enzymes applied to isolate the protoplast and the tissue material should be placed in proper osmolyticum or plasmolyticum. Generally mannitol, a sugar alcohol at high concentration, an easily transportable component provides a good stable osmotic environment which prevents bursting of protoplast.

Protoplasts may be isolated by any one of the two following ways:

- Mechanical method (Non-enzymatic method);
- Enzymatic method.

Mechanical Method

Any soft parenchymatous tissue is kept in a plasmolyticum. The plasmolysed tissue is then finely chopped into pieces and the intact cells (plasmolysed) are released into the medium from the cut surface. The suspension is then allowed for deplasmolysis and the released protoplasts attain their original size. The yield of protoplast in this method is very low, for large scale of protoplast yield the enzymatic method is followed.

Enzymatic Method

Young fully expanded soft leaves, or in vitro grown callus tissue or cell suspension culture grown cells can be used as the source material. The tissues or cells are incubated in plasmolyticum for 1 hr before enzymatic treatment. The intact tissue materials cut into smaller pieces to increase the surface area of enzymatic activity. The enzymes can be used either sequentially in two step method or in a single step by mixed enzymatic method.

The enzymes used are of three main categories:

- Cellulases:

  Cellulysin;

  Driselase.
- Hemicellulase:

  Helicase;

  Rhozyme.
- Pectinase:

  Macerozyme-R-10;

  Pectinol.

The concentration of enzymes used and the time period of incubation varies greatly depending on the tissue type. In two step method, the pectinase and hemicellulases are applied first and then cellulase is applied for complete removal of cell wall.

After release of protoplast into the suspension, for removal of enzymes the protoplasts are collected in centrifuge tube as pellet and washed several times with the osmoticum.

Purification of Protoplasts

For purification, the protoplasts suspended in osmoticum are centrifuged using sucrose (20%) solution. The viable protoplasts float on the top surface of sucrose solution forming a band. These protoplasts are then collected, re-suspended in osmoticum and washed several times.

Finally the protoplasts suspended in a measured volume of protoplast culture medium after counting the number with the help of hemocytometer. The viability of protoplast is checked with the help of Fluorescin diacetate staining or phenosafranine or calcofluore white.

Isolation of Sub-Protoplasts

Sub-protoplasts do not contain the entire content of plant cells, these can be of different types such as cytoplasts, karyoplasts or mini-protoplasts or micro-protoplasts or microplasts.

- Cytoplasts lack the nucleus and contain the entire cytoplasm of a cell or part of it. These are often used for cybrid production.
- Karyoplasts or Mini-protoplasts contain a nucleus surrounded by some cytoplasm and the original outer plasma membrane.
- Micro-protoplasts contain fraction of both nuclear and cytoplasmic material.
- Microplasts lack the nuclear material and contain only fraction of cytoplasm and outer membrane.

Protoplast Viability Test

The most frequently used staining methods for checking protoplast viability are:

- Fluorescein diacetate (FDA) dissolved in acetone is used at a conc. of 0.01% and intact viable protoplasts only fluoresce when observed under UV.
- Phenosafranine is also used at a conc. of 0.01 %, which is specific for dead protoplast that shows red in colour.

Somatic Cell Hybridization

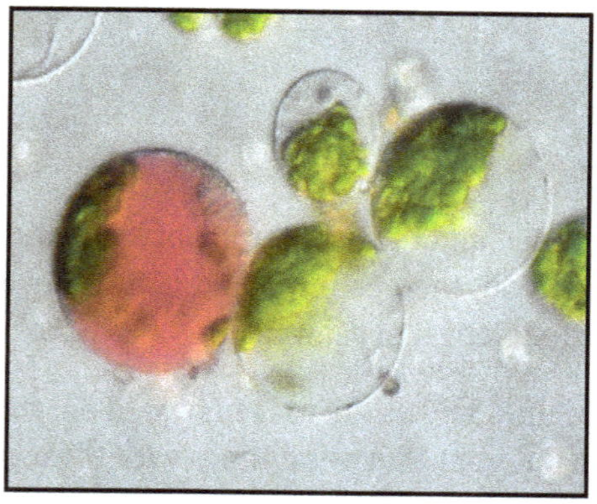

The process by which protoplasts of two different plant species fuse together to form hybrids is known as somatic hybridisation and the hybrids so produced is known as somatic hybrids.

Sexual hybridization since time immemorial has been used as a method for crop improvement but it has its own limitations as it can only be used within members of same species or closely related wild species. Thus, this limits the use of sexual hybridization as a means of producing better varieties. Development of viable cell hybrids by somatic hybridization, therefore, has been considered as an alternative approach for the production of superior hybrids overcoming the species barrier. The technique can facilitate breeding and gene transfer, bypassing problems associated with conventional sexual crossing such as, interspecific, intergeneric incompatibility. This technique of hybrid production via protoplast fusion allows combining somatic cells (whole or partial) from different cultivars, species or genera resulting in novel genetic combinations including symmetric somatic hybrids, asymmetric somatic hybrids or somatic cybrids.

The most common target using somatic hybridization is the gene of symmetric hybrids that contain the complete nuclear genomes along with cytoplasmic organelles of both parents. This is unlike sexual reproduction in which organelle genomes are generally contributed by the maternal parent. On the other hand, somatic cybridization is the process of combining the nuclear genome of one parent with the mitochondrial and/or chloroplast genome of a second parent. Cybrids can be produced by donor-recipient method or by cytoplast-protoplast fusion. Incomplete asymmetric somatic hybridization also provides opportunities for transfer of fragments of the nuclear genome, including one or more intact chromosomes from one parent (donor) into the intact genome of a second parent (recipient).

These methods provide genetic manipulation of plants overcoming hurdle of sexual incompatibility, thereby, serving as a method of bringing together beneficial traits from taxonomically distinct species which cannot be achieved by sexual crosses. Several parameters such as, source tissue, culture medium and environmental factors influence ability of a protoplast derived hybrid cells to develop into a fertile plant. The general steps involved in somatic hybridization and cybridization methods are elaborated in figures below.

Steps Involved in Somatic Hybridization

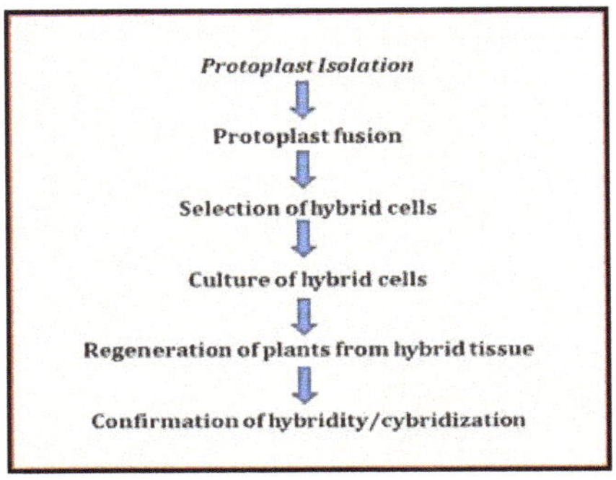

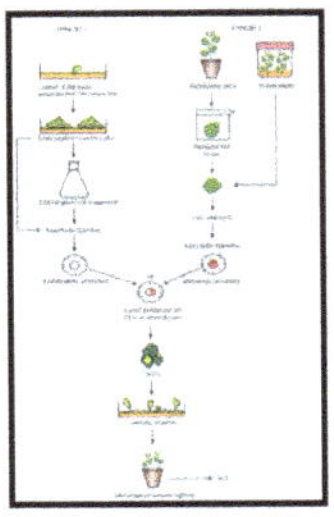

Figure: Schematic view of symmetric protoplast fusion producing somatic hybrids.

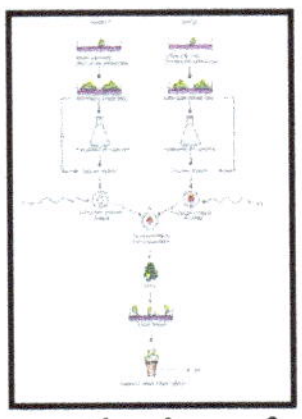

Figure: Schematic view of asymmetric protoplast fusion using donor-recipient method resulting into creation of alloplasmic somatic hybrid or cybrids.

Protoplast Fusion

Protoplast fusion could be spontaneous during isolation of protoplast or it can be induced by mechanical, chemical and physical means. During spontaneous process, the adjacent protoplasts fuse together as a result of enzymatic degradation of cell walls forming homokaryons or homokaryocytes, each with two to several nuclei. The occurrence of multinucleate fusion bodies is more frequent when the protoplasts are prepared from actively dividing callus cells or suspension cultures. Since the somatic hybridization or cybridization require fusion of protoplasts of different origin, the spontaneous fusion has no value. To achieve induced fusion, a suitable chemical agent (fusogen) like, NaNO3, high Ca2+, polyethylene glycol (PEG), or electric stimulus is needed.

i. Fusion by means of NaNO3: It was first demonstrated by Kuster in 1909 that the hypotonic solution of NaNO3 induces fusion of isolated protoplast forming heterokaryon (hybrid). This method was fully described by Evans and Cocking, however this method has a limitation of generating few no of hybrids, especially when highly vacuolated mesophyll protoplasts are involved.

ii. High pH and Ca++ treatment: This technique lead to the development of intra- and interspecific hybrids. It was demonstrated by Keller and Melcher in 1973. The isolated protoplasts from two plant species are incubated in 0.4 M mannitol solution containing high Ca++(50 mM CaCl2.2H2O) with highly alkaline pH of 10.5 at 37°C for about 30 min. Aggregation of protoplasts takes place at once and fusion occurs within 10 min.

iii. Polyethylene glycol treatment: Polyethylene glycol (PEG) is the most popularly known fusogen due to ability of forming high frequency, binucleate heterokaryons with low cytotoxicity. With PEG the aggregation occurred mostly between two to three protoplasts unlike Ca++ induced fusion which involves large clump formation. The freshly isolated protoplasts from two selected parents are mixed in appropriate proportions and treated with 15-45% PEG (1500-6000MW) solution for 15-30 min followed by gradual washing of the protoplasts to remove PEG. Protoplast fusion occurs during washing. The washing medium may be alkaline (pH 9-10) and contain a high Ca++ ion concentration (50 mM). This combined approach of PEG and Ca++ is much more efficient than the either of the treatment alone. PEG is negatively charged and may bind to cation like Ca++, which in turn, may bind to the negatively charged molecules present in plasma lemma, they can also bind to cationic molecules of plasma membrane. During the washing process, PEG molecules may pull out the plasma lemma components bound to them. This would disturb plamalemma organization and may lead to the fusion of protoplasts located close to each other. The technique is nonselective thus, induce fusion between any two or more protoplasts.

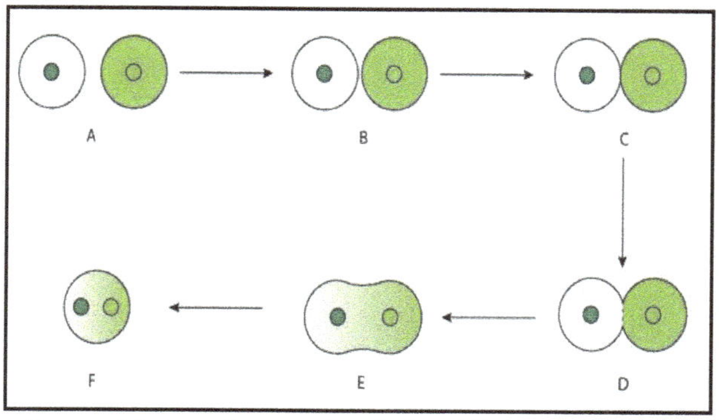

Figure: Sequential stages in protoplast fusion. (A) two separate protoplasts, (B) agglutination of two protoplasts, (C and D) Membrane fusion at localized site, and (E and F) development of spherical heterokaryon.

iv. Electrofusion: The chemical fusion of plant protoplast has many disadvantages – (1) The fusogen are toxic to some cell systems, (2) it produces random, multiple cell aggregates, and (3) must be removed before culture. Compare to this, electrofusion is rapid, simple, synchronous and more easily controlled. Moreover, the somatic hybrids produced by this method show much higher fertility than those produced by PEG-induced fusion.

Zimmermann and Scheurich demonstrated that batches of protoplasts could be fused by electric fields by devising a protocol which is now widely used. This protocol involves a two-step process. First, the protoplasts are introduced into a small fusion chamber containing parallel wires or plates which serve as electrodes. Second, a low-voltage and rapidly oscillating AC field is applied, which causes protoplasts to become aligned into chains of cells between electrodes. This creates complete cell-to-cell contact within a few minutes. Once alignment is complete, the fusion is induced by application of a brief spell of high-voltage DC pulses (0.125-1 kVcm-1). A high voltage DC pulses induces a reversible breakdown of the plasma membrane at the site of cell contact, leading to fusion and consequent membrane reorganization. The entire process can be completed within 15 min.

Selection of Fusion Products

The somatic hybridization by electrofusion of protoplasts allow one-to-one fusion of desired pairs of protoplasts and, therefore, it is easy to know the fate of fusion products. However, protoplast suspension recovered after chemical treatments (fusogen) consists of the following cell types:

- Unfused protoplasts of the two species/strains,
- Products of fusion between two or more protoplasts of the same species (homokaryons), and
- 'Hybrid' protoplasts produced by fusion between one (or more) protoplasts of each of the two species (heterokaryons).

The heterokaryons which are the potential source of future hybrids constitute of a very small (0.5-10%) proportion of the mixture. Therefore, an effective strategy has to be employed for their identification and isolation. Various protocols have been proposed and practiced for the effective selection of hybrids, including morphological basis, complementation of biochemical and genetic traits of the fusing partners, and manual or electronic sorting of heterokaryons/hybrid cells.

Morpho-physiological basis: The whole mixtures of the protoplasts are cultured after fusion treatment and the resulting calli or regenerants are screened for their hybrid characteristics. Occasionally the hybrid calli outgrow the parental cell colonies and are identified by their intermediate morphology, i.e. green with purple coloured cells. However, the process is labour intensive and requires glasshouse facilities. It is limited to certain combinations showing differences in their regeneration potential under specific culture conditions.

Complementation: In this case complementation or genetic or metabolic deficiencies of the two fusion partners are utilized to select the hybrid component. When protoplasts of two parents, (one parent bearing cytoplasmic albino trait and the other parent bearing green trait) each parent carrying a non-allelic genetic or metabolic defect are fused, it reconstitutes a viable hybrid cell, of wild type in which both defects are mutually abolished by complementation, and the hybrid cells are able to grow on minimal medium non-permissive to the growth of the parental cells bearing green trait. Later, the calli of hybrid nature could be easily distinguished from the parental type tissue (albino trait) by their green color. The complementation selection can also be applied to dominant characters, such as dominant resistance to antibiotics, herbicides or amino acid analogues.

Isolation of heterokaryons or hybrid cells: The manual or electronic isolation of heterokaryons or hybrid cells is the most reliable method. Manual isolation requires that the two parental type protoplasts have distinct morphological markers and are easily distinguishable. For example, green vacuolated, mesophyll protoplasts from one parent and richly cytoplasmic, non green protoplasts from cultured cells of another parent. The dual fluorescence method also helps easy identification of fusion products. In this case, the protoplast labeled green by treatment with fluorescein diacetate (FDA, 1-20 mgl-1) are fused with protoplasts emitting a red fluorescence, either from chlorophyll autofluorescence or from exogenously applied rhodamine isothiocyanate (10-20 mgl-1). The labeling can be achieved by adding the compound into the enzyme mixture. This can be applied even for morphologically indistinguishable protoplasts from two parents.

Verification and Characterization of Somatic Hybrids

As no system is foolproof and they have their own advantages and disadvantages. Therefore, even after selecting the desired hybrids/cybrids following protoplast fusion, it is required to carry out one or more tests to compare the parent protoplast lines with the putative hybrids. Some of the techniques that can be tried are:

Morphology: Somatic hybrids in most of the cases show characters intermediate between the two parents such as, shape of leaves, pigmentation of corolla, plant height, root morphology and other vegetative and floral characters. The method is not much accurate as tissue culture conditions may also alter some morphological characters or the hybrid may show entirely new traits not shown by any of the parents.

Isozyme analysis: Multiple molecular forms of same enzyme which catalyses similar or identical reactions are known as isozymes. Electrophoresis is performed to study banding pattern as a check for hybridity. If the two parents exhibit different band patterns for a specific isozyme the putative hybrid can be easily verified. The isozymes commonly used for hybrid identification include, acid phosphatase, esterase, peroxidase.

Cytological analysis: Chromosome counting of the hybrid is an easier and reliable method to ensure hybridity as it also provides the information of ploidy level. Cytologically the chromosome count of the hybrid should be sum of number of chromosomes from both the parents. Besides number of chromosomes, the size and structure of chromosomes can also be monitored. However, the approach is not applicable to all species, particularly where fusion involves closely related species or where the chromosomes are very small. Moreover, sometimes the somaclonal variations may also give rise to different chromosome number.

Molecular analysis: Specific restriction pattern of nuclear, mitochondrial and chloroplast DNA characterizes the plastomes of hybrids and cybrids. Molecular markers such as RFLP, RAPD, ISSR can be employed to detect variation and similarity in banding pattern of fused protoplasts to verify hybrid and cybrid.

Cybrids or Cytoplasmic Hybrids

Sexual hybridization involves fusion of the nuclear genes of both the parents but somatic hybrids involves even cytoplasm from both the parental species in hybrid obtained by protoplast fusion. However, in another case somatic hybrids containing nuclear genome of one parent but cytoplasm from both the parents, are termed as cybrids. The approach is time consuming and require several years of crossing plants provides an opportunity to study interparental mitochondrial, chloroplast fusion giving rise to plants with novel genomes.

Methods to Produce Cybrids

They are produced in variable frequencies in normal protoplast fusion experiments due to one of the following methods:

1. Fusion of normal protoplast with an enucleated protoplast. The enucleated protoplast can be produced by high speed centrifugation (20,000-40,000xg) for 60 min with 5-50% percoll.
2. Fusion between a normal protoplast and another protoplast with a non-viable nucleus or suppressed nucleus.
3. Elimination of one of the nuclei after heterokaryons formation.
4. Selective elimination of chromosomes at a later stage.
5. Irradiating (with X-rays or gamma rays) the protoplasts of one species prior to fusion in order to inactivate their nuclei.
6. By preparing enucleate protoplasts (cytoplasts) of one species and fusing them with normal protoplasts of the other species.

Cybrids provide the following unique opportunities: (i) transfer of plasmogenes of one species into the nuclear background of another species in a single generation, and even in (ii) sexually incompatible combinations, (iii) recovery of recombinants between the parental mitochondrial or chloroplast DNAs (genomes), and (iv) production of a wide variety of combinations of the parental and recombinant chloroplasts with the parental or recombinant mitochondria.

Applications of Somatic Hybridization

1. Novel interspecific and intergeneric crosses which are difficult to produce by conventional methods can be easily obtained.
2. Important characters, such as resistance to diseases, ability to undergo abiotic stress and other quality characters, can be obtained in hybrid plant by the fusion of protoplasts of plant bearing particular character to the other plant which may be susceptible to diseases.
3. Protoplasts of sexually sterile haploid, triploid, aneuploid plants can be fused to obtain fertile diploids and polyploids.
4. Studying cytoplasmic genes may be helpful to carry out plant breeding.

5. Most of the agronomically important traits, such as cytoplasmic male sterility, antibiotic resistance and herbicide resistance, are cytoplasmically encoded, hence can be easily transferred to other plant.
6. Plants in juvenile stage can also be hybridized by means of somatic hybridization.
7. Somatic hybridization can be used as a method for the production of autotetraploids.

Limitations of Somatic Hybridization

1. Application of protoplast methodology requires efficient plant regeneration system from isolated protoplasts. Protoplasts from two species can be fused, however, production of somatic hybrids is not easy.
2. Lack of a proper selection method for fused products (hybrids) poses a problem.
3. The end product of somatic hybridization are often unbalanced (sterile, misformed and unstable).
4. Somatic hybridization of two diploids leads to formation of amphidiploids which is unfavorable.
5. It is not sure for a character to completely express after somatic hybridization.
6. The regeneration products of somatic hybridization are often variable due to somaclonal variation, chromosome elimination, organelle segregation.
7. All diverse intergeneric somatic hybrids are sterile and, therefore, have limited chances of development of new varieties.
8. To transfer useful genes from wild species to cultivated crop, it is necessary to achieve intergeneric recombination or chromosome substitution between parental genomes.

Gene Gun Method

The gene gun method is a method used for genetically modifying plants.

With the use of a gene gun, the gene gun method delivers extra DNA directly into a plant's nucleus. The method is also commonly called particle acceleration or microprojectile bombardment.

The gene gun can be used on seedlings or tissue culture cells. Prior to injecting the DNA into the plant tissue via the gene gun, either microscopic gold or tungsten particles are liberally coated with many hundreds of copies of genes. The particles are then forced into the nucleus with the gene gun.

The first gene gun method of DNA modification was developed in 1980. The process relied on a 22-caliber cartridge to shoot the gene-coated metal particles into the plant's cell.

Nowadays, modern versions of the gene gun operate differently. The modernized gene gun depends on a vacuum chamber and high-pressure gas to hurl the metal particles into the plant cells.

With each firing of the gene gun, the pressurized gas is released in a sudden popping burst and the gene coated metal particles fire into the cells.

The gene gun is highly effective at modifying the DNA and genetics of plant cells.

Genetic modification in plants can make them resistant to drought, disease, and pests. It also helps create a more nutritious edible plant with higher levels of proteins.

This method was used to create several commercial events such as RoundupReady soybean, it has several limitations that make it less appealling than the use of Agrobacterium as a transformation method.

The gene gun is a good example of a creative idea being developed into a practical technology.

The gene gun can be used on either tissue culture cells or seedlings (to make chimeric plants) of any species. As the name implies, this method works by shooting DNA into the plant cells.

Gene Gun Design

The gene gun was originally a Crosman air pistol modified to fire dense tungsten particles. It was invented by John C Sanford, Ed Wolf and Nelson Allen at Cornell University, and Ted Klein of DuPont, between 1983 and 1986. The original target was onions (chosen for their large cell size) and it was used to deliver particles coated with a marker gene. Genetic transformation was then proven when the onion tissue expressed the gene.

The earliest custom manufactured gene guns (fabricated by Nelson Allen) used a 22 caliber nail gun cartridge to propel an extruded polyethylene cylinder (bullet) down a 22 cal. Douglas barrel. A droplet of the tungsten powder and genetic material was placed on the bullet and shot down the barrel at a lexan"stopping" disk with a petri dish below. The bullet welded to the disk and the genetic material blasted into the sample in the dish with a doughnut effect (devastation in the middle, a ring of good transformation and little around the edge). The gun was connected to a vacuum pump and was under vacuum while firing. The early design was put into limited production by a Rumsey-Loomis (a local machine shop then at Mecklenburg Rd in Ithaca, NY, USA). Later the design was refined by removing the "surge tank" and changing to nonexplosive propellants. DuPont added a plastic extrusion to the exterior to visually improve the machine for mass production to the scientific community. Bioradcontracted with Dupont to manufacture and distribute the device. Improvements include the use of helium propellant and a multi-disk-collision delivery mechanism. Other heavy metals such as gold and silver are also used. Gold may be favored because it has better uniformity than tungsten and tungsten can be toxic to cells, but its use may be limited due to availability and cost.

Biolistic Construct Design

A construct is a piece of DNA inserted into the target's genome, including parts that are intended to be removed later. All biolistic transformations require a construct to proceed and while there is great variation among biolistic constructs, they can be broadly sorted into two categories: those which are designed to transform eukaryotic nuclei, and those designed to transform prokaryotic-type genomes such as mitochondria, plasmids or plastids.

Those meant to transform prokaryotic genomes generally have the gene or genes of interest, at least one promoter and terminator sequence, and a reporter gene; which is a gene used to enable detection or removal of those cells which didn't integrate the construct into their DNA. These genes may each have their own promoter and terminator, or be grouped to produce multiple gene products from one transcript, in which case binding sites for translational machinery should be placed between each to ensure maximum translational efficiency. In any case the entire construct is flanked by regions called border sequences which are similar in sequence to locations within the genome, this allows the construct to target itself to a specific point in the existing genome.

Constructs meant for integration into a eukaryotic nucleus follow a similar pattern except that: the construct contains no border sequences because the sequence rearrangement that prokaryotic constructs rely on rarely occurs in eukaryotes; and each gene contained within the construct must be expressed by its own copy of a promoter and terminator sequence.

Though the above designs are generally followed, there are exceptions. For example, the construct might include a Cre-Lox system to selectively remove inserted genes; or a prokaryotic construct may insert itself downstream of a promoter, allowing the inserted genes to be governed by a promoter already in place and eliminating the need for one to be included in the construct.

Application

Gene guns are so far mostly used with plant cells. However, there is much potential use in humans and other animals as well.

Plants

The target of a gene gun is often a callus of undifferentiated plant cells or a group of immature embryos growing on gel medium in a Petri dish. After the DNA-coated gold particles have been delivered to the cells, the DNA is used as a template for transcription (transient expression) and sometimes it integrates into a plant chromosome ('stable' transformation).

If the delivered DNA construct contains a selectable marker, then stably transformed cells can be selected and cultured using tissue culture methods. For example, if the delivered DNA construct contains a gene that confers resistance to an antibiotic or herbicide, then stably transformed cells may be selected by including that antibiotic or herbicide in the tissue culture media.

Transformed cells can be treated with a series of plant hormones, such as auxins and gibberellins, and each may divide and differentiate into the organized, specialized, tissue cells of an entire plant. This capability of total re-generation is called totipotency. The new plant that originated from a successfully transformed cell may have new traits that are heritable. The use of the gene gun may be contrasted with the use of Agrobacterium tumefaciens and its Ti plasmid to insert DNA into plant cells.

Advantages

Biolistics has proven to be a versatile method of genetic modification and it is generally preferred to engineer transformation-resistant crops, such as cereals. Notably, Bt maize is a product of biolistics. Plastid transformation has also seen great success with particle bombardment when compared to other current techniques, such as Agrobacterium mediated transformation, which have difficulty targeting the vector to and stably expressing in the chloroplast. In addition, there are no reports of a chloroplast silencing a transgene inserted with a gene gun. Additionally, with only one firing of a gene gun, a skilled technician can generate two transformed organisms. This technology has even allowed for modification of specific tissues in situ, although this is likely to damage large numbers of cells and transform only some, rather than all, cells of the tissue.

Limitations

Biolistics introduces DNA randomly into the target cells. Thus the DNA may be transformed into whatever genomes are present in the cell, be they nuclear, mitochondrial, plasmid or any others, in any combination, though proper construct design may mitigate this. Another issue is that the gene inserted may be overexpressed when the construct is inserted multiple times in either the same or different locations of the genome. This is due to the ability of the constructs to give and take genetic material from other constructs, causing some to carry no transgene and others to carry multiple copies; the number of copies inserted depends on both how many copies of the transgene an inserted construct has, and how many were inserted. Also, because eukaryotic constructs rely on illegitimate recombination, a process by which the transgene is integrated into the genome without similar genetic sequences, and not homologous recombination, which inserts at similar sequences, they cannot be targeted to specific locations within the genome, unless the transgene is co-delivered with genome editing reagents.

Electroporation

Electroporation is a popular physical method for introducing new genes directly into the protoplasts.

In this method, electric field is playing important role. Due to the electric field protoplast get temporarily permeable to DNA. In electroporation, plant cell protoplasts are kept in an ionic solution containing the vector DNA in a small chamber that has electrodes at opposite ends. A pulse of high voltage is applied to the electrode which makes the transient pores (ca. 30 nm) in the plasma membrane, allowing the DNA to diffuse into the cell. Immediately, the membrane reseals.

If appropriately treated, the cells can regenerate cell wall, divide to form callus and, finally, regenerate complete plants in suitable medium. The critical part of the procedure is to determine conditions which produce pores that are sufficiently large and remain open long enough to allow for DNA diffusion. At the same time, the conditions should make pores that are temporary. With a 1 cm gap between the electrodes and protoplasts of 40-44µm diameter, 1-1.5 kVcm -2 of field strength for 10µs is required for efficient introduction of DNA. It was seen that presence of 13% PEG (added after DNA) during electroporation significantly raised the transformation frequency. The other factors which may improve the transformation frequency by electroporation are linearizing of plasmid, use of carrier DNA, and heat shock (45 ~ for 5 min) prior to addition of vector, and placing on ice after pulsing. Under optimal conditions transformation frequencies of up to 2% have been reported. Stably transformed cell lines and full plants of a number of cereals have been produced through electroporation.

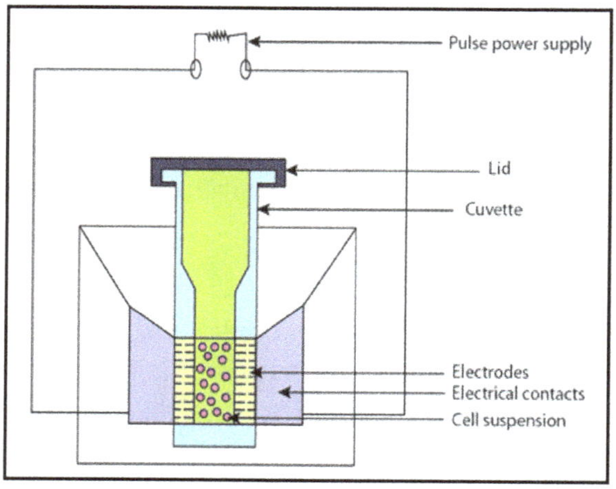

Figure: Electroporation.

There are some parameters that can be considered when performing in vitro electroporation:

Cell size

Cell size is inversely correlated to the size of the external field needed to generate permeabilization. Consequently, optimization for each cell type is essential. Likewise, cell orientation matters for cells that are not spherical.

Temperature

It has been observed that plant membrane resealing is effectively temperature dependent and shows slow closure at low temperatures. For DNA transfer, it has been found that cooling at the time of permeabilization and subsequent heating in incubator increases transfer efficacy and cell viability.

Post-pulse Manipulation

Cells are susceptible when in the permeabilized state, and it has been shown that waiting for 15min after electroporation in order to allow resealing before pipetting cells, increases cell viability.

Composition of Electrodes and Pulsing Medium

For short pulses is needed for release of metal from the standard aluminium electrodes used in standard disposable cuvettes. Some authors advocate the use of low conductivity or more resistance media for DNA transfer in order to increase viability and increase transfection efficacy.

Microinjection

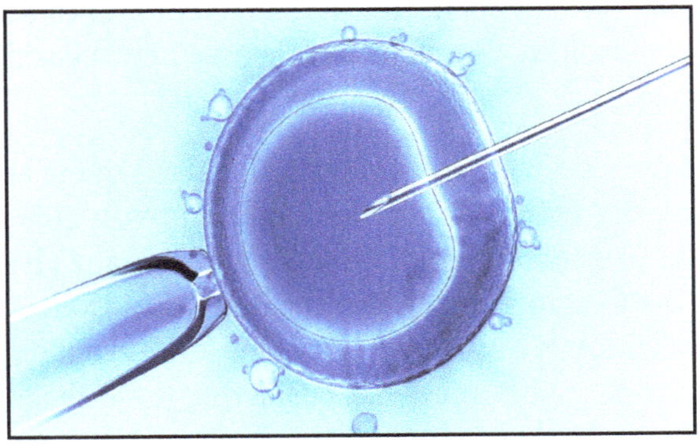

Microinjection is an established method to introduce DNA into single cells to generate transient and stable transformants. The injection of single cells within plant organs, however, is more difficult, and only a low percentage of the injected cells survive. The Eppendorf micromanipulation system has been used to inject tomato, mustard and Arabidopsis seedlings.

Usually plant cells possess a very rigid cell wall, which is a problem for microinjection since it requires stable injection capillaries with small diameters. In combination with the cellular turgor, the wall is responsible for the structural integrity of a plant seedling. The cell turgor of plants can be very high and is not only responsible for the morphology of a seedling but also plays an important role in development. Even if the injection against the high internal pressure was successful, the injury of a larger cell wall area will automatically lead to the destruction of the cell, at the latest when the needle is retreated.

The large central vacuole of a plant cell is also a problem that diminishes the chances of a successful cytoplasmic injection. The cytoplasm surrounding the central vacuole is only a few micrometres thick. As a result of the deformation of the cell wall during the injection and the sudden relaxation when the capillary penetrates the wall, the needle will often end up in the vacuole. Material injected into the vacuole is not only trapped but in many cases degraded, due to the low pH and the activity of several proteases. In addition, the tonoplast – the single membrane surrounding the vacuole – is easily injured, which leads to leakage of cytotoxic substances into the cytoplasm and hence to cell death.

Furthermore, because of multiple cell layers, light scattering and pigmentation, the optical properties of whole plant seedlings or plant organs are not optimal. Therefore, it is often difficult to identify different cell layers or even subcellular structures like vacuoles under microscopic control during the microinjection process.

To approach the described problems, Eppendorf developed a micromanipulation system equipped with very fast micromotors, variable injection parameters, high holding and injection pressure, in combination with a piezo step function. The set-up was tested for its suitability to inject plant cells — specifically tomato, mustard and Arabidopsis seedlings.

The micromanipulation system's injection angle was adjusted to 35° and axial injection movements were performed at a speed of 3000 µm/s with a step size of 10–20 µm. These settings ensure minimal injuries of the cell wall and thus raise the survival rate after injection. The injection movement can be supported by synchronised piezo impulses when the system's InjectMan 4 and PiezoXpert devices are electronically connected. As a marker for injection, Lucifer yellow (ly) — a water-soluble dye that easily allows discrimination between cytoplasmic and vacuolar localised fluorescence — was injected.

Typically, 5–8 injections per seedling were performed into cells of the etiolated hypocotyl. In the case of the tomato and mustard seedling injections, efficiencies of 26% and 48% were obtained. The injection efficiency for Arabidopsis seedlings was about 10% at 24 h after injection — strongly decreased compared to the tomato and mustard seedlings. However, Arabidopsis seedlings are much smaller and very fragile. As a consequence, injected cells do not only die relatively often but in many cases the whole hypocotyl is affected. Considering these challenges, 10% injection efficiency reflects a good value for Arabidopsis seedlings in our opinion.

Localised fluorescence of Iy 24 h after injection indicated that the cells were viable, but ly is no in vivo marker. Hence, we decided to inject Alexa Fluor 488-labelled histone (histone-AF488). After injection into the cytoplasm, the histones should be transported into the nucleus, where they accumulate. Histone import into the nucleus is an active process that requires GTP and intact protein complexes. Therefore, the import of histone-AF488 should only occur in living cells. The percentage of cells showing histone-AF488 fluorescence was in the same range as the observed percentage after ly injection, confirming that these cells survived the injection procedure.

The Eppendorf micromanipulation system, with its fast and precise motor movements, was thus well suited for the injection of plant cells. This includes the injection of Arabidopsis seedlings with relatively high efficiencies. Since a large collection of Arabidopsis mutants and transgenic lines is available for many research areas, microinjection of Arabidopsis cells from specific organs and tissues should, in addition to injection of other plant species, help to analyse numerous scientific questions.

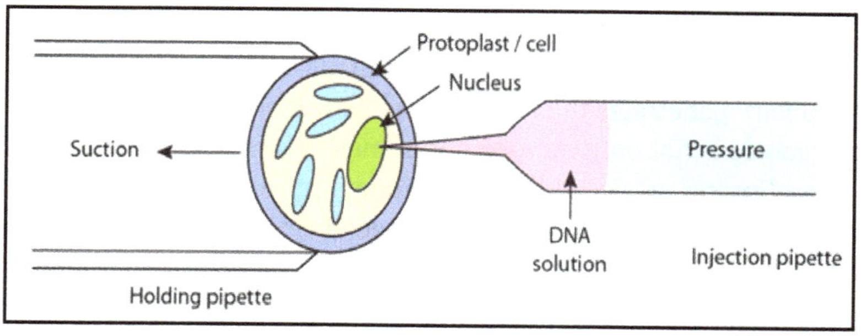

Agrobacterium

Agrobacterium elicits neoplastic growths (called crown gall tumors) that affect most dicotyledonous plants. Moreover, although plants represent the natural hosts for Agrobacterium, this microorganism can also genetically transform a wide range of other eukaryotic species, from yeast to mushrooms and filamentous fungi to phytopathogenic fungi to human cells. Most functions for Agrobacterium-host cell DNA transfer are coded by a large (200-kb) tumor-inducing (Ti) plasmid that resides in the bacterial cell and carries two important genetic components: the vir (virulence) region and the T-DNA delimited by two 25-bp direct repeats at its ends, termed the T-DNA borders. The vir region comprises seven major loci, virA, virB, virC, virD, virE, virG, and virH, which encode most of the bacterial protein machinery (Vir proteins) of the DNA transport. After induction of vir gene expression by small phenolic signal molecules secreted from wounded susceptible plant cells, the T-DNA borders are nicked by the bacterial VirD2 endonuclease, generating a transferable single-stranded (ss) copy of the bottom strand of the T-DNA region, designated the T strand.

Interestingly, the T strand does not travel alone but is thought to directly associate with two Agrobacterium proteins, VirD2 and VirE2, forming a transport (T) complex in which one molecule of VirD2 is covalently attached to the 5′-end of the T strand, whereas VirE2, an ssDNA-binding protein, is presumed to cooperatively coat the rest of the T strand molecule. Although only the wild-type T-DNA carries Ti genes, any DNA placed between the T-DNA borders will be transferred to the plant host. This lack of sequence specificity implies that a T-DNA molecule itself does not encode protein machinery for its transport from the bacterial cell into the host cell, import into the host cell nucleus, and integration into the host cell genome. Instead, these functions are fulfilled by the bacterial Vir proteins and their host cell partners.

In the last quarter of a century, since the discovery of the stable integration of the bacterial DNA in crown gall tumors, Agrobacterium has served as a primary a tool for plant genetic engineering. Furthermore, the Agrobacterium-host cell interaction also represents a unique and powerful experimental system to study a wide spectrum of basic biological processes such as cell-cell recognition and cell-to-cell transport, nuclear import, assembly and disassembly of protein-DNA complexes, DNA recombination, and regulation of gene expression.

Genetic Engineering

Predating the science fiction visions of Nano machines performing genetic engineering and other biotechnological tasks, Agrobacterium is a present-day microscopic but extremely complex machine routinely used to alter genotypes of higher plants. This use of Agrobacterium is based on its unique capacity for "trans-kingdom sex", i.e. transfer of genetic material between prokaryotic and eukaryotic cells. Decades of research altered, augmented, and vastly improved this natural capacity, resulting in ingeniously modified Agrobacterium strains that can transfer and stably integrate virtually any gene to a variety of plant species, from research model plants such as Arabidopsis to agriculturally important rice and corn. In this issue, Valentine reviews the molecular mechanisms of T-DNA transfer that underlie the use of Agrobacterium as gene vector. Importantly, a special emphasis is made on the ethical and political aspects of the release of Agrobacterium-generated genetically modified plants into the environment and on the attitudes of different and often opposing geopolitical and economic forces to these matters.

Cell-Cell Recognition and Attachment

Agrobacterium recognition of and attachment to the host cells is an early and essential step of the infection process. The bacterial proteins participating in these events are encoded by several loci, e.g. chvA, chvB, pscA, and att. In particular, the att genes, such as attR and attD, are located on a cryptic plasmid pAtC58. Although pAtC58 is about twice as large as the Ti plasmid and, thus, likely encodes numerous bacterial functions, it has been considered dispensable for infection. In this issue, Nair et al. revisit this dogma and demonstrate that pAtC58 increases Agrobacterium virulence. Surprisingly, however, this increase was likely due to the enhancement of the vir gene expression rather than to the presence of the attR gene on the pATC58 plasmid.

Although the bacterial genes participating in Agrobacterium attachment to the host plants are relatively well characterized, the involvement of host factors remains largely obscure. Recent progress in this direction stemmed from identification of T-DNA-tagged Arabidopsis mutants defective in their ability to bind Agrobacterium. In this issue, Zhu et al. (2003) report characterization of one such mutant, rat4, which contains a T-DNA insertion in the 3′-untranslated region of cellulose synthase-like gene CSLA9. Their data indicate that CSLA9 is involved in development and growth of lateral roots, determination of sugar composition of plant cell walls, and the ability of the roots to bind Agrobacterium. Furthermore, the CSLA9 promoter exhibited enhanced expression in the root elongation, previously shown to be most susceptible to Agrobacterium-mediated transformation. Similarly, preferential expression in the elongation zones of Arabidopsis roots has been reported for a plant histone gene, H2A-1, which is required for T-DNA integration.

Cell-to-Cell Transport

T strands with their cognate VirD2 are exported into the host cell by a type IV secretion system, which in Agrobacterium is assembled from proteins encoded by the virD4 gene and virB operon, with 11 open reading frames. Interestingly, this system also exports other Agrobacterium proteins, such as VirE3, VirF, and VirE2. Although VirE2 most likely packages the T strand into the T complex, this binding is thought to occur within the cytoplasm of the host cell after independent export of the T strands and VirE2. But what prevents VirE2 from binding to the T strands already within

Agrobacterium? This role has been assigned to the VirE1 chaperone protein that associates with VirE2 and blocks its binding to ssDNA. Furthermore, VirE1 has been suggested also to participate in the VirE2 export. However, convincing evidence to the contrary is presented by Vergunst. Using the Cre Recombinase Reporter Assay for Translocation to study protein export from Agrobacterium into plant and yeast cells, they demonstrate that recognition of VirE2 by the bacterial export machinery and its subsequent translocation into host cells does not depend on the presence of VirE1. Cre Recombinase Reporter Assay for Translocation was then utilized to show that VirE3, another product of the virE locus earlier reported to be exported into the yeast cells, is transferred from Agrobacterium to plants. The function of VirE3 in the plant cell, however, remains completely unknown. Vergunst also localized the Agrobacterium-to-plant export signal to the C-terminal 50 amino acids of VirE2 and VirE3. These and previous studies clearly demonstrate that in addition to DNA, Agrobacterium transfers a variety of its own proteins to the host cell; the challenge to researchers now is to understand how these bacterial proteins participate in the genetic transformation process from within the host cell.

Dna Integration and Expression

T-DNA integration is the culmination point of the entire process of the Agrobacterium-plant cell DNA transfer. But how does a T-DNA molecule insinuate itself into the molecule of the plant genomic DNA? Because the T-DNA does not encode enzymatic activities required for integration, the protein components of the T-complex, i.e. VirD2 and VirE2 and plant DNA ligases, must provide these functions. T-DNA integration was proposed to initiate with ligation of the 5′ end of the T strand to the genomic DNA followed by second strand synthesis by the plant DNA repair machinery. However, another study suggested that the T strand is converted into a double-stranded form before integration.

Obviously, it is the expression of the integrated transgenes that produces tumors in the wild-type infection or desired transgenic phenotypes in genetic engineering experiments. However, some transgenes, although stably integrated in the plant genome, often are not expressed due to their posttranscriptional gene silencing (PTGS), which is characterized by a reduction in transcript levels without affecting the rate of transcription. Furthermore, PTGS of a transgene can also silence its endogenous cellular homologs, resulting in co-suppression. Agrobacterium-mediated gene delivery has served as a valuable experimental tool to study the mechanisms of induction and cell-to-cell spread of PTGS. Continuing this trend, Lee utilized silencing of Agrobacterium oncogenes contained within the T-DNA to demonstrate that transgene sequences influence the effectiveness of PTGS and that sequences required for oncogene silencing must include a translation start site. Unexpectedly, unlike several other cases of PTGS, oncogene silencing was not graft transmissible. In addition to helping us better understand the mechanisms of PTGS, silencing of Agrobacterium oncogenes represents a novel approach to control the crown gall disease, which affects such agronomically important plants as grape, rose, apple, cherry, and others.

Finally, Agrobacterium infection was also used as a model system to study cellular processes required for establishment of plant tumors. Wächter showed that vascular differentiation and disruption of epidermal cell layers play key roles in tumor formation, allowing delivery of water and Suc to the proliferating tumor cells. This work suggested that Agrobacterium-induced tumors represent nutritional sinks to which a high-volume flow of solutes with essential inorganic nutrients is directed in an ethylene-dependent fashion, followed by a Suc-to-hexose shift in sugar balance caused by vacuolar invertases within the growing tumors.

Examples of Genetic Modification

Bananas

In many countries around the world bananas are the main source of calories. According to reports from Uganda, their production is compromised by the emergence of new diseases. Ugandan scientists have successfully used a genetic modification, inserting a pepper gene into bananas, which prevents the fruit from getting the disease.

Onions that do not Make you Cry

In 2008, a New Zealand research team lead by Colin Eady produced an onion that does not make you cry while cutting it. Interestingly, the insertion of a single gene which down regulates the activity of the onion enzyme that makes your eyes water has managed to achieve two things: firstly, onions no longer make your eyes water, and secondly, they now have even more health beneficial sulphur-containing substances than regular onions.

Golden Rice

On 31st July 2000, Ingo Potrykus appeared on the cover of Time magazine. The Swiss scientist and his German colleague Peter Beyerhad had produced a breed of rice which, unlike any other, also contains provitamin A. The lack of this vitamin is especially harmful to the poorest and is estimated to cause blindness among 250,000-500,000 children every year. Another two million people a year die from other deficiency-related causes. So far, the measures taken to introduce vitamin supplements have not yet reached those poor countries. Because of its colour, the product was first given the name golden rice, which remains the same to this day. After concluding numerous tests, researchers from the International Rice Research Institute in the Philippines have proven that even small amounts of the rice are sufficient, and that it is absolutely safe. Unfortunately, partly due to the vandalism of "green" activists, the rice has yet to reach its target group. We should also mention the existence of genetically modified rice, produced in 2011, which contains four times as much iron as the regular one, and could therefore save even more lives.

Purple Tomatoes

In 2008, a small English research group published a study describing how they had transferred a gene from a decorative plant into a tomato, which enabled the production of anthocyanin, making the tomato dark blue. They later tested it on mice, and discovered that it prevents them from getting cancer. These tomatoes cannot be bought due to their GMO status but you can buy similar dark blue ones, which are the result of a complex interspecific hybridisation performed by Italian scientists. Producing such tomatoes seems to have sparked competition between cultivators. In 2012 for instance, the Israelis announced that they have beaten the Italians by introducing a new species to the market called Black Galaxy.

Carrots that Help Prevent Osteoporosis

In 2004, an American research team transferred a CAX1 gene from the mouseear cress into carrots so that they contained larger amounts of organically bound calcium. In 2008, they performed a study where such carrots were tested on mice and 30 volunteers, and the results showed that humans absorbed 42% more calcium from the modified carrots than from regular ones. The aim of this test was to help prevent osteoporosis, while the emphasis was on its bioavailability in target tissues.

Soybean Oil for Frying

Two American companies have significantly improved soybean edible oil through genetic modification. The Plenish oil marketed by DuPont used gene silencing to produce oil that contains low levels of polyunsaturated fats and high levels of monounsaturated fats, while the saturated fatty acids have decreased by 20%. This kind of oil has high stability during baking, meaning it does not need to go through the chemical process of hydrogenation, which produces unwanted trans fats. There is a similar vascular-system-friendly oil marketed by Monsanto with two inserted genes enabling the production of omega-3 fatty acids. Rapeseed is also close to being released into production by the company BASF, containing five genes isolated from seaweed. These are the first products since the Flavr Savr tomato in 1994 to be made directly available to the consumer.

Arctic Apple

An apple turns brown if it is cut in half, which is why its slices are often soaked in antioxidants to prevent this. In 2012, Canadian scientists started releasing two popular apple cultivars, Golden Delicious and Granny Smith, and are planning to introduce Gala and Fuji cultivars, all under the brand name Arctic Apple. The non-browning characteristic could already be found in some apple cultivars, like the newest Slovene cultivar called Majda, but none of them has gained strong commercial recognition. The aim of this research is to prevent the most popular cultivars from turning brown. The same principle applies to these apples as to the previously mentioned onions: it is not about adding genes but about downregulating the activity of the existing ones.

References

- G., Halford, Nigel (2012). Genetically modified crops. World Scientific (Firm) (2nd ed.). London: Imperial College Press. ISBN 978-1848168381. OCLC 785724094
- Transgenic-plants-meaning-reasons-and-fundamentals-588: biotechnologynotes.com, Retrieved 12 March 2020
- Bruening, G.; Lyons, J. M. (2000). "The case of the FLAVR SAVR tomato". California Agriculture. 54 (4): 6–7. doi:10.3733/ca.v054n04p6
- Cisgenics-and-its-Application-in-Crop-Improvement-3861: biotecharticles.com, Retrieved 30 May 2020
- Andrew Pollack (19 June 2013). "Executive at Monsanto wins global food honor". The New York Times. Retrieved 20 June 2020
- Gene-gun-method-3145: maximumyield.com, Retrieved 20 June 2020
- Koornneef, M.; Meinke, D. (2010). "The development of Arabidopsis as a model plant". The Plant Journal. 61 (6): 909–21. doi:10.1111/j.1365-313X.2009.04086.x. PMID 20409266
- Microinjection-into-plant-cells-of-etiolated-seedlings-981244058: labonline.com.au, Retrieved 20 April 2020
- 10-successful-examples-of-genetic-modification: metinalista.si, Retrieved 25 June 2020
- Predieri, S. (2001). "Mutation induction and tissue culture in improving fruits". Plant Cell, Tissue and Organ Culture. 64 (2/3): 185–210. doi:10.1023/A:1010623203554

Permissions

All chapters in this book are published with permission under the Creative Commons Attribution Share Alike License or equivalent. Every chapter published in this book has been scrutinized by our experts. Their significance has been extensively debated. The topics covered herein carry significant information for a comprehensive understanding. They may even be implemented as practical applications or may be referred to as a beginning point for further studies.

We would like to thank the editorial team for lending their expertise to make the book truly unique. They have played a crucial role in the development of this book. Without their invaluable contributions this book wouldn't have been possible. They have made vital efforts to compile up to date information on the varied aspects of this subject to make this book a valuable addition to the collection of many professionals and students.

This book was conceptualized with the vision of imparting up-to-date and integrated information in this field. To ensure the same, a matchless editorial board was set up. Every individual on the board went through rigorous rounds of assessment to prove their worth. After which they invested a large part of their time researching and compiling the most relevant data for our readers.

The editorial board has been involved in producing this book since its inception. They have spent rigorous hours researching and exploring the diverse topics which have resulted in the successful publishing of this book. They have passed on their knowledge of decades through this book. To expedite this challenging task, the publisher supported the team at every step. A small team of assistant editors was also appointed to further simplify the editing procedure and attain best results for the readers.

Apart from the editorial board, the designing team has also invested a significant amount of their time in understanding the subject and creating the most relevant covers. They scrutinized every image to scout for the most suitable representation of the subject and create an appropriate cover for the book.

The publishing team has been an ardent support to the editorial, designing and production team. Their endless efforts to recruit the best for this project, has resulted in the accomplishment of this book. They are a veteran in the field of academics and their pool of knowledge is as vast as their experience in printing. Their expertise and guidance has proved useful at every step. Their uncompromising quality standards have made this book an exceptional effort. Their encouragement from time to time has been an inspiration for everyone.

The publisher and the editorial board hope that this book will prove to be a valuable piece of knowledge for students, practitioners and scholars across the globe.

Index

www.ingramcontent.com/pod-product-compliance
Lightning Source LLC
Chambersburg PA
CBHW081212180926
51613CB00003B/217